Hellwich · Siebert

Stereochemistry Workbook

T0225710

K.-H. Hellwich · C. D. Siebert

Stereochemistry
Workbook

191 Problems and Solutions

translated by Allan D. Dunn

 Springer

Authors

Dr. Karl-Heinz Hellwich
Postfach 100731
63007 Offenbach
Germany
e-mail: *khellwich@web.de*

Dr. Carsten D. Siebert
65936 Frankfurt am Main
Germany
e-mail: *dr.cdsiebert@web.de*

Translator

Dr. Allan D. Dunn
65824 Schwalbach/Ts.
Germany

Library of Congress Control Number: 2006930686

DOI 10.1007/978-3-540-32912-9

ISBN 978-3-540-32911-4 Springer Verlag Berlin Heidelberg New York

ISBN 978-3-540-32912-9 (eBook)

Springer is a part of Springer Science+Business Media

springer.com
© Springer-Verlag Berlin Heidelberg 2006

Typesetting: ptp, Berlin
Production: LE-TeX Jelonek, Schmidt & Vöckler GbR, Leipzig
Cover design: KünkelLopka Werbeagentur, Heidelberg

Printed on acid-free paper   45/3100/YL - 5 4 3 2 1 0

Dr. phil. nat. Karl-Heinz Hellwich

was born in 1962 and studied chemistry with particular emphasis on stereochemistry at the University of Frankfurt am Main, Germany. During his doctoral studies 1989–1995 which were centred on drugs for the regulation of lipid metabolism at the Institute of Pharmaceutical Chemistry of the above university he also taught organic chemistry. In addition, in 1991–2001 he gave lectures on chemical nomenclature and on stereochemistry for pharmacy students at the Universities of Frankfurt a. M. and Jena, Germany. In 1993 he became an external referee for and in 1998 a member of the IUPAC Commission on Nomenclature of Organic Chemistry. After publishing a well accepted text book and several translations of specialist publications, he joined the Beilstein Chemiedaten und Software GmbH in 1999 and then the Beilstein-Institut in Frankfurt a. M. where he has been reviewing and editing data for inclusion in the Beilstein Database. Since 2006 he has been a Titular Member of the Division Committee of the IUPAC Division on Chemical Nomenclature and Structure Representation.

Dr. phil. nat. Carsten D. Siebert

was born in 1967 and studied chemistry and pharmacology at the University of Frankfurt a. M., Germany. His doctoral studies at the Institute of Pharmaceutical Chemistry were centred on the synthesis and testing of neuroprotectants and were carried out in collaboration with Merck KGaA and the Universities of Vienna, Austria, and Berlin, Germany. During this time he also taught organic chemistry and stereochemistry to pharmacy students. In 1999 he joined the Beilstein Chemiedaten und Software GmbH where he reviewed and edited data for inclusion in the Beilstein Database. In 2001 he moved to ABDATA Pharma-Daten-Service a business group of the Werbe- und Vertriebsgesellschaft Deutscher Apotheker mbH. ABDATA publishes pharmaceutical, chemical and medical data for health service professionals in pharmacies, hospitals and surgeries. In addition to the supervision of comprehensive monographs for currently prescribed drugs, he is the editor responsible for a computer-based warning system widely used in German pharmacies to alert pharmacists to allergic reactions of drugs. He is considered as an authority on the stereochemistry of drugs and collaborates regularly with authors of pharmacological textbooks.

Dedicated to Prof. Dr. Hermann Linde (1929–2001)

Foreword

As the author of what was probably the first modern textbook on stereo-chemistry (Eliel, "Stereochemistry of Carbon Compounds" McGraw-Hill, 1962) and the co-author or "godfather" of three recent texts on the subject (refs. 1–3 in the Appendix/Bibliography) it gives me great pleasure that this comprehensive problem book on the subject has finally appeared in English. Many times over the last 44 years I have been asked where students could find exercise problems to help with the study of the above texts, and the answer has always been that the teachers would have to make up their own problem sets. No longer! Having made up many such problem sets myself, I am well aware that they have been nowhere as extensive (nor usually as comprehensively stimulating) in covering the subject matter as this book is.

The 191 problems in this book cover most of the area of stereochem-istry, including nomenclature, stereogenic elements (centers, axes, planes) and their descriptors, symmetry, inorganic stereochemistry, determination of enantiomer excess, conformation of acyclic and cyclic compounds, and more. The answers, in addition to providing solutions to the problems, fre-quently include additional explanations of the underlying principles. The problems are ordered more or less in order of increasing difficulty. (I had a hard time with some of the problems toward the end myself!)

A number of the questions asked relate to natural and/or pharmaceutical products. This should help stimulate and maintain the interest of future pharmacists, pharmacologists and physicians to study these problems, and not just that of future chemists and molecular biologists.

I mentioned that this book relieves teachers of making up their own problems. But this makes the role of the teacher in no way redundant. First, since the problems are not keyed to any particular book, they may not be in the order in which the subject matter is presented in a course; so they need to be assigned as the course proceeds. By the same token the total number of problems is probably too large for most students to handle, so a selection in the assignment might be advisable.

The answers to many of the problems can and should give rise to stimulating discussions. A possible way to handle this also, in view of the fact, that all the answers are in the book and thus there is no point for them to be graded may be to place the students in discussion groups and let them argue over the answers (preferably with sets of simple, inexpensive molecular models). In the end the teacher may have to enter these discussions to resolve the most difficult questions! Which is also a good way to judge the students' thinking and reasoning power.

And now I invite you to dig into the problems!

Ernest L. Eliel

University of North Carolina at Chapel Hill
July 2006

Table of Contents

Introduction

The idea for this workbook was born out of experience. During several years of assistantship in teaching organic chemistry to pharmacy students and subsequent employment in editing and publishing chemical and pharmaceutical information, the problem of unambiguous descriptions for three-dimensional structures of chemical compounds often arose and it became apparent that it is not enough just to acquire mere textbook knowledge but it is of particular importance to be able to reproduce spatially correct stereoformulae using examples of actual compounds. In this stereochemistry workbook the authors draw on their wealth of knowledge to meet this demand. Since the book is not intended as a substitute for a textbook, the reader is recommended to refer to the literature cited in the appendix for detailed accounts of all the various aspects of stereochemistry.

The book starts by asking the reader to define some basic terms used in stereochemistry. The answers listed in the second part of the book not only give the solutions to the problems, but in addition provide some of the tools required to tackle subsequent problems which are arranged in increasing order of difficulty and complexity. After the questions on definitions mentioned above, there are some simple exercises to determine relative and absolute configuration and to recognise their significance. It soon becomes apparent to the reader that it is still not common practice to view molecules in three dimensions, even although the basic principles have been known since the 19th century. Next the reader is introduced to more complicated problems involving stereoselective and stereospecific chemical reactions and the determination of symmetry point groups. The answers to all these questions are accompanied with precise representations of chemical structures and in depth explanations to assist and train the reader to approach each new problem with a high degree of preparedness for more challenging problems.

The authors have carefully selected their examples not only to provide practice in addressing a broad range of stereochemical problems and analysing stereochemical relevant reactions but which nearly always are also representative of actual pharmaceutical compounds and molecules of biochemical importance. Since receptors and enzymes in most cases show stereoselective recognition of ligand and substrate, respectively, to understand the effect of a compound on an organism the precise spatial description of the pharmaceutical/medicinal agent employed must obviously be taken into account. It is unfortunately, that even nowadays the demands for the complete characterisation of molecules are often not met either by chemists or by pharmacologists despite the fact that compounds can be synthesised essentially free of isomers and that mixtures of stereoisomers can be separated into their components. However, many drugs with chirality centres are marketed as mixtures of enantiomers or diastereomers. In many such cases one isomer is ineffective but sometimes the undesired isomer is the reason for adverse side effects of the finished drug and the organism must, in most instances, eliminate an unnecessary xenobiotic. Thus, it remains a mystery why many pharmaceutical textbooks still neglect stereochemistry and the stereochemical representation of drugs, although there are numerous stereochemistry textbooks available on the subject. As a result, there are often incomplete and hence incorrect descriptions of molecules in the former.

The broad range of potential applications illustrated clearly demonstrates that this is not niche science and that an appreciation of stereochemistry is a fundamental requirement for a profound understanding of biological processes. For this reason this book is strongly recommended for molecular pharmacologists and physicians in addition to students of natural sciences. The authors hope that their choice of compounds in the exercises in this book will reflect the interdisciplinary nature and importance of stereochemistry.

Questions

❓ 1

Explain briefly (in one or two short sentences) the meaning of the following basic stereochemical terms.

a) chirality
b) constitution
c) configuration
d) conformation
e) stereoisomers

❓ 2

Draw formulae of all possible isomers which have the empirical formula C_3H_6O.

❓ 3

What is meant by the term absolute configuration and how is it specified?

❓ 4

Determine the configuration of the isomer of 2-hydroxysuccinic acid shown below.

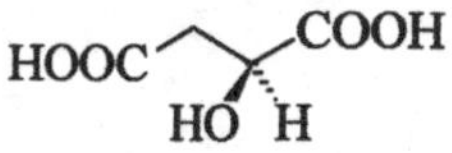

❓ 5

Define clearly the following terms. In each case give one example to illustrate your answer.

a) enantiomerism
b) diastereomers
c) racemate
d) epimer

❓ 6

Which of the following properties or methods can be used to distinguish between (R)-carvone and (S)-carvone?

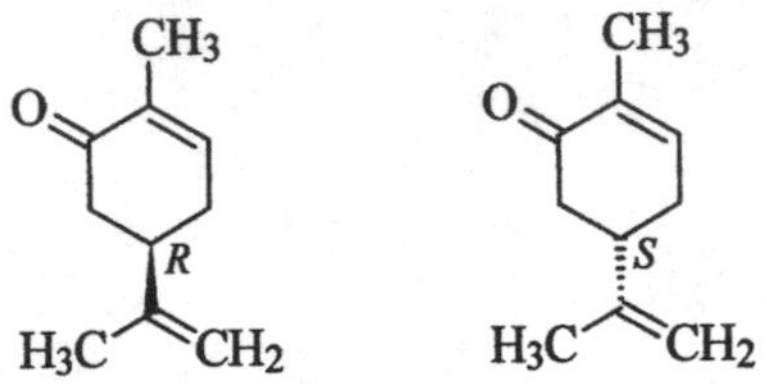

a) boiling point
b) UV spectroscopy
c) refractive index
d) melting point
e) smell
f) optical rotation
g) dipole moment
h) circular dichroism
i) NMR spectroscopy
j) IR spectroscopy

❓ 7

Determine the configuration of the isomer of the amino acid alanine shown below.

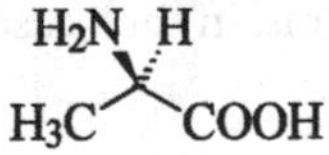

❓ 8

What is meant by the terms
a) symmetry element,
b) meso compound?

? 9

Mark all the chirality centres in the formula of the lipid-lowering drug lovastatin shown below with an asterisk (*). How many chirality centres are present?

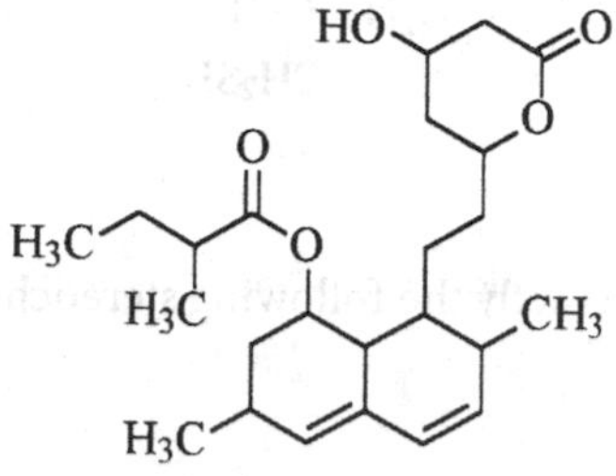

? 10

Explain clearly and succinctly the following terms.
a) mutarotation
b) enantioselective
c) retention
d) stereogenic unit

? 11

When is a molecule chiral?

? 12

Explain briefly and unambiguously the meaning of the following terms.
a) atropisomers
b) anomers
Which stereodescriptors are used to describe them?

? 13

Define the following terms.
a) stereoselectivity
b) stereospecific

❓ 14

Deduce the absolute configuration of L-cysteine according to the R/S nomenclature.

$$
\begin{array}{c}
\text{COOH} \\
\text{H}_2\text{N} \!\!-\!\!\!\!-\!\!\text{H} \\
\text{CH}_2\text{SH}
\end{array}
$$

❓ 15

Explain clearly and succinctly the following stereochemical terms.
a) inversion
b) prochiral
c) topicity

❓ 16

How many prochirality centres has butanone? Where are they in the molecule?

❓ 17

Explain in a few short sentences what is meant by the term relative configuration. Which stereodescriptors can be used to describe the relative configuration?

❓ 18

Draw as Newman projections the different conformations of ethylene glycol ($HO-CH_2-CH_2-OH$) and label each clearly.

❓ 19

What is a pseudochirality centre?

❓ 20

What is the difference between the descriptor pairs Re/Si and re/si? Which of these two stereodescriptor pairs can be used to describe the two faces of the planar part of the structures of the following compounds?

a) (*R*)-3-chlorobutan-2-one
b) (2*R*,4*S*)-2,4-dichloropentan-3-one

❓ 21

The psychostimulant adrafinil is a racemate. Draw the structural formulae of both compounds.

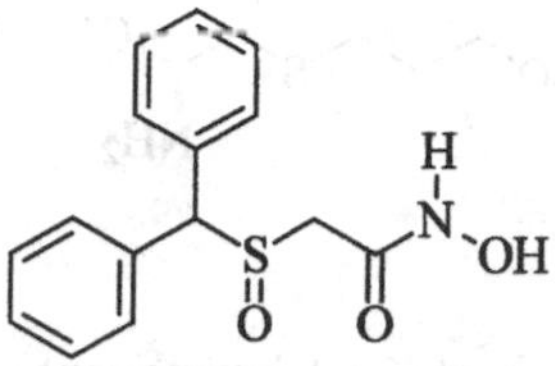

❓ 22

The chromatographic purification of 1 g of (−)-ethyl lactate with an enantiomeric excess (ee) of 85 % yields, without any loss of material, the optically pure (−)-enantiomer. How many g of the (+)-enantiomer were separated?

❓ 23

Draw the formulae of all the possible isomeric butenes and determine their symmetry elements and point groups. Use the flow chart in the appendix to assist you.

❓ 24

Draw the formulae of all possible isomers of 2-methylcyclohexan-1-ol. What relationship do these isomers have to one another?

❓ 25

The antiseptic debropol is used as the racemate. Which enantiomer is shown below?

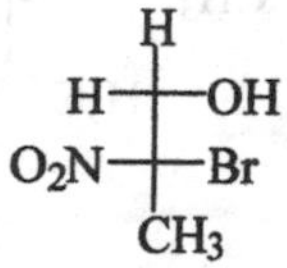

❓ 26

Draw the structural formula of (Z)-2-cyano-3,4-dimethylpent-2-enoic acid methyl ester.

❓ 27

Which enantiomer of the mucolytic fudosteine is shown below?

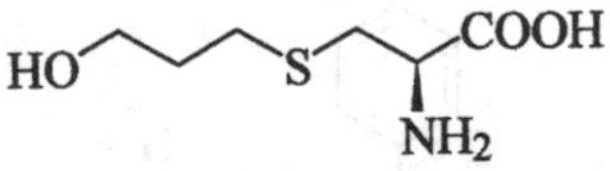

❓ 28

Draw the structural formula of (Z)-1-bromopenta-1,2,3-triene.

❓ 29

Which of the following isomers differ in constitution and which in configuration?
a) (E)-1-bromopropene and (Z)-1-bromopropene
b L-alanine and β-alanine
c) lactic acid and 3-hydroxybutanoic acid
d) (−)-lactic acid and (+)-lactic acid
e) 1-chloropropene and 2-chloropropene
f) cis-2-chlorocyclohexanol and trans-2-chlorocyclohexanol

❓ 30

Determine the configuration of the double bonds in the cytostatic retinoid alitretinoin. How many stereoisomers are possible?

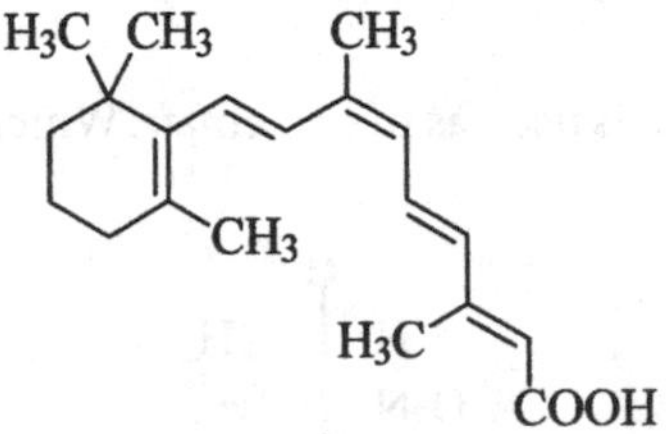

❓ 31

How many configurational isomers are there correlated with the constitution expressed in the following names? In those cases where two isomers exist state their relationship to each other.

a) ethanol
b) butan-2-ol
c) glycerol
d) 2,3-dibromobutane
e) acetone oxime
f) pent-3-en-2-ol
g) pentane-2,3-diol
h) pentane-2,4-diol
i) 3-bromobutan-2-ol
j) but-2-enoic acid
k) 4-ethylhepta-2,5-diene
l) hexa-2,3,4-triene

❓ 32

Convert the formula of galactose shown below into a Fischer projection formula and state whether it is the α or the β anomer.

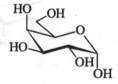

❓ 33

Are there compounds with a constitution where

a) enantiomers but no diastereomers are possible,
b) both enantiomers as well as diastereomers exist,
c) diastereomers but no enantiomers are possible?

Give examples where appropriate.

❓ 34

Draw the formulae of all possible isomers of difluorocyclobutane and determine their symmetry elements and point groups. Assume that the cyclobutane ring is planar. Indicate which isomers are chiral. Use the flow chart in the appendix to assist you.

❓ 35

Draw the structural formulae of all the epimers of (2R,3S)-bicyclo[2.2.1]-heptane-2,3-diol.

❓ 36

Determine the absolute configuration of the following compounds.

a)

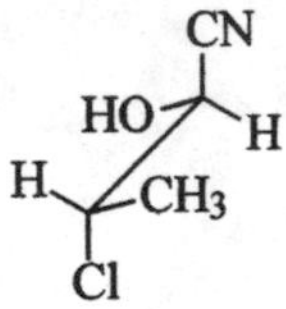

b)

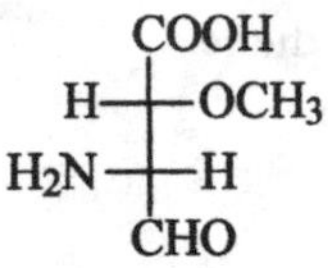

$$
\begin{array}{c}
\text{COOH} \\
\text{H}\!-\!\!|\!-\!\text{OCH}_3 \\
\text{H}_2\text{N}\!-\!\!|\!-\!\text{H} \\
\text{CHO}
\end{array}
$$

c)

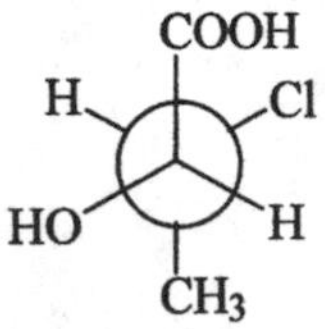

d)

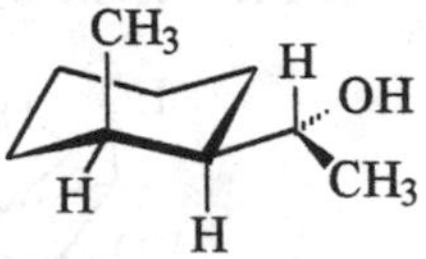

❓ 37

What is the relationship between the following pairs of compounds?

a)

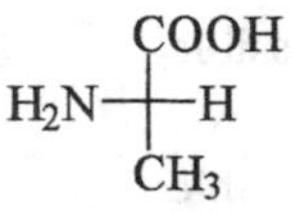

$$
\begin{array}{c}
\text{COOH} \\
\text{H}_2\text{N}\!-\!\!|\!-\!\text{H} \\
\text{CH}_3
\end{array}
\quad \text{and} \quad
\begin{array}{c}
\text{NH}_2 \\
\text{HOOC}\diagup\!\!\diagdown\text{CH}_3 \\
\text{H}
\end{array}
$$

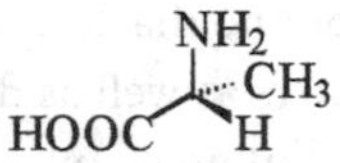

b)

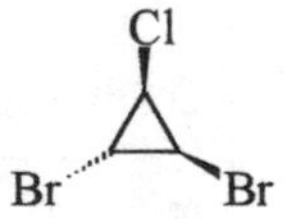

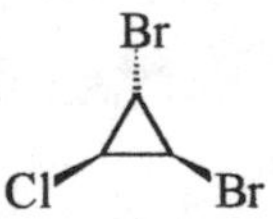

c)

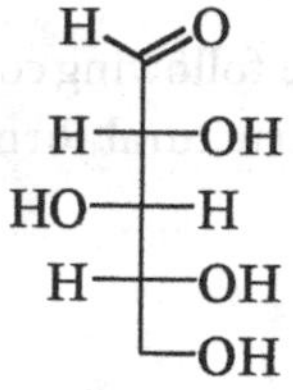

and

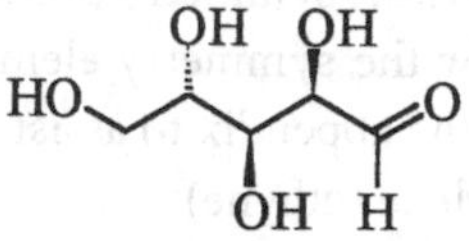

d)

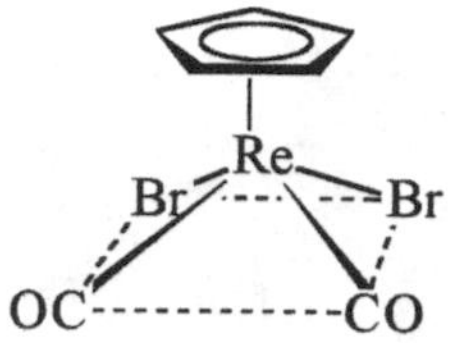

and

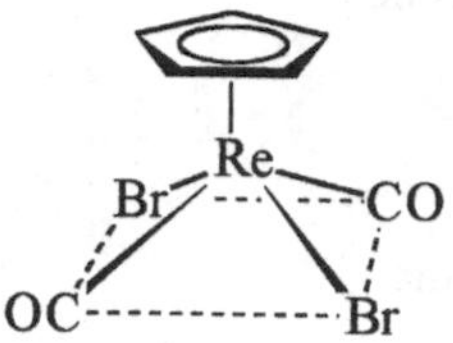

❓ 38

Are the following compounds enantiomers or diastereomers?

a) (E)-1,2-dichloroethene and (Z)-1,2-dichloroethene
b) (+)-tartaric acid and *meso*-tartaric acid
c) (1R,2S)-cyclohexane-1,2-diamine and (1R,2R)-cyclohexane-1,2-diamine
d) (1S,2S)-cyclohexane-1,2-diamine and (1R,2R)-cyclohexane-1,2-diamine
e) α-D-glucopyranose and β-D-glucopyranose
f) α-D-mannopyranose and α-L-mannopyranose

❓ 39

In order to determine the absolute configuration or the enantiomeric excess of a compound containing a hydroxy group, it is often esterified with a pure enantiomer of Mosher's acid (3,3,3-trifluoro-2-methoxy-2-phenylpropanoic acid, MTPA). Which configuration has the ester derived from (S)-1-phenyl-propan-1-ol and (R)-Mosher's acid chloride?

40

Determine the symmetry point groups of the following compounds. In each case show the symmetry elements in the structural formula. Use the flow chart in the appendix to assist you.
a) acetylene (ethyne)
b) hydrogen peroxide (H_2O_2)
c) white phosphorus (P_4)
d) ferrocene

e) twistane

41

What is the absolute configuration of the analgesic drug vedaclidine?

42

The antibiotic linezolid is the pure S enantiomer. Draw the structural formula of the molecule with this configuration.

❓ 43

Give precise names for the following compounds.

a)

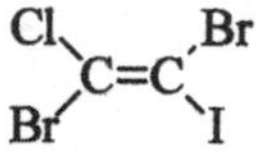

b)

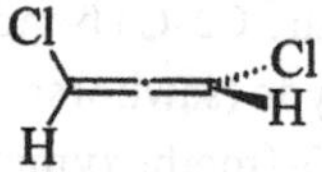

c)

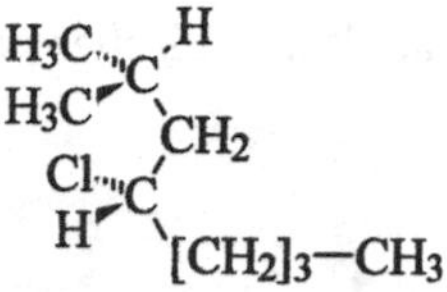

d)

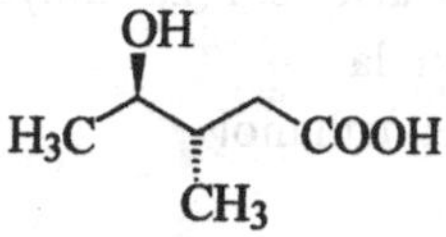

❓ 44

What relationship do the following pairs of compounds have to each other?

a)

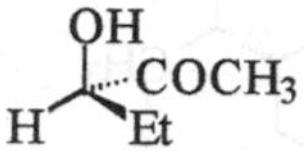

and

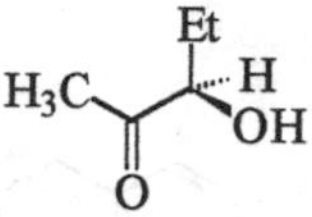

b)

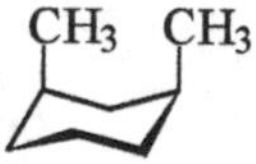

and

c)

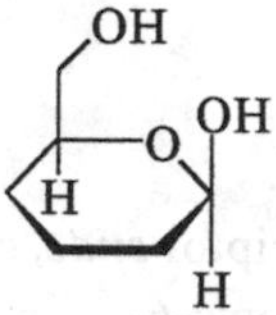

and

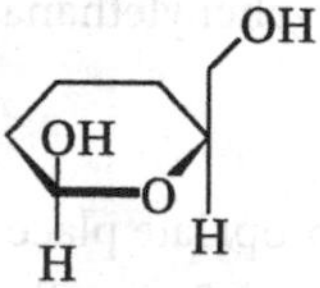

d)

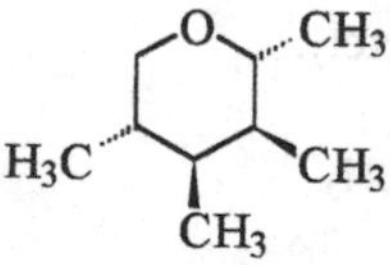

and

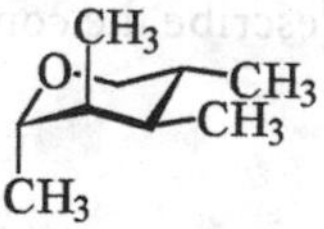

❓ 45

Draw unambiguous structural formulae for the following compounds.

a) L-*erythro*-2-amino-3-hydroxybutanoic acid (as a Newman projection viewed along the C2-C3 bond)

b) D-glyceraldehyde (sawhorse projection)

c) (*Z*)-4-bromo-3-(methoxymethyl)but-2-enoyl chloride

d) *u*-3-bromopentan-2-ol (zigzag projection)

e) (*R,R*)-tartaric acid (2,3-dihydroxybutanedioic acid) as a Fischer projection formula

f) (*S*)-(1-^{2}H$_1$)ethanol

❓ 46

Nateglinide is an orally administered antidiabetic. Assign suitable stereodescriptors to describe the structure and determine the amino acid from which it is derived.

❓ 47

Describe a non-chromatographic method for the separation of the enantiomers of *rac*-1-phenylethanamine.

❓ 48

Insert at the appropriate place a stereodescriptor *endo, exo, syn* or *anti* into the systematic name 7-.....-ethyl-5-.....-isopropyl-6-.....-methyl-7-.....-propyl-bicyclo[2.2.1]hept-2-ene of the compound shown below. Explain why this name cannot describe the compound completely.

❓ 49

How many isomers are there of diamminedichloridoplatinum(II), $[PtCl_2NH_3)_2]$? Determine the symmetry elements and point groups for all isomers and assign appropriate stereodescriptors.

❓ 50

How many stereoisomers are there of 1,4-dimethylbicyclo[2.2.1]heptan-2-ol (theoretical possible number and actual number)?

❓ 51

Draw the structural formula of *l*-1,2-dichlorocyclobutane.

❓ 52

The potassium channel activator cromakalim is a mixture of *trans*-configured compounds. What relationship do these two stereoisomers have to each other?

❓ 53

What relationship do the following pairs of compounds have to each other?

a)

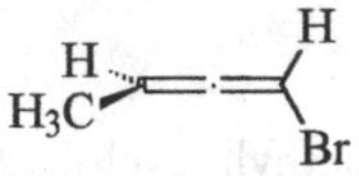

and

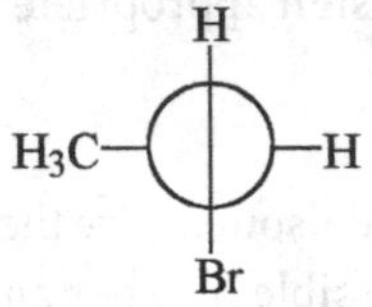

b)

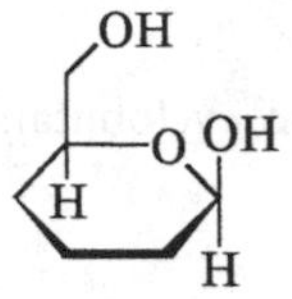

and

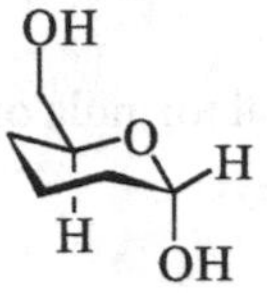

c)

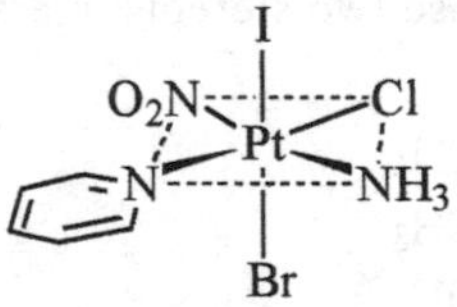

and

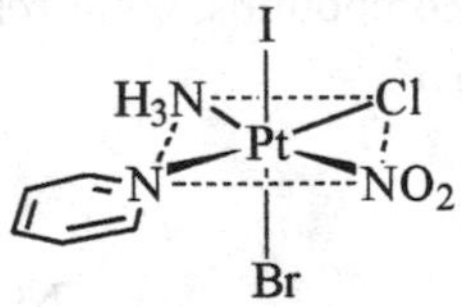

d)

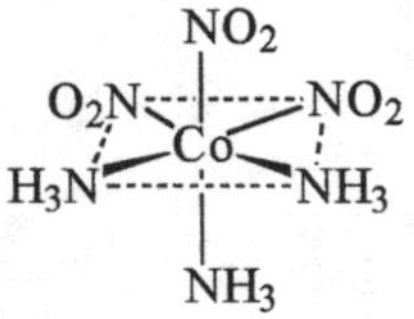

and

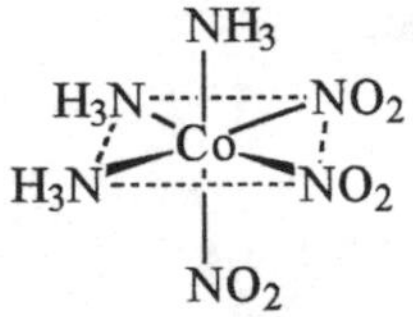

❓ 54

Describe precisely and completely the configuration of the following compound. How many stereoisomers are there of the compound?

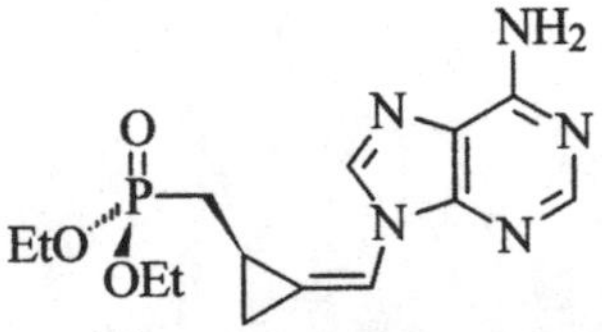

❓ 55

Draw the structural formula of 1-bromo-4t-chloro-4-methylcyclohexane-1r-carboxylic acid.

❓ 56

Draw the structural formula of (R)-1-bromobuta-1,2-diene.

❓ 57

Deduce the symmetry point groups of all the isomers of $[CrCl_2(NH_3)_4]^+$ and assign a precise stereodescriptor for each isomer.

❓ 58

The diuretic cyclothiazide is used as a mixture of stereoisomers. How many stereoisomers can exist?

❓ 59

What other stereodescriptor can be used to describe the configuration of (R_a)-1,3-dichloroallene unambiguously?

❓ 60

The proton pump inhibitor nepaprazole is a racemate. Draw the structural formulae of both u-configured isomers.

❓ 61

Are the two faces of the double bonds in the following compounds homotopic, enantiotopic or diastereotopic? Give, where possible, a suitable descriptor for the face oriented towards you.

a)

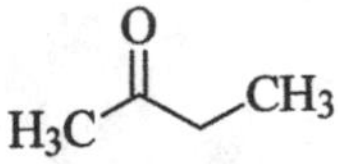

b)

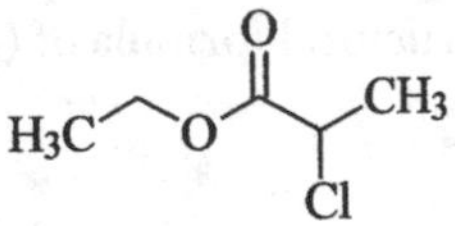

c)

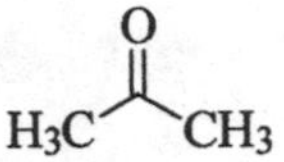

d)

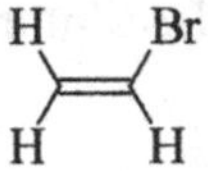

❓ 62

What products are formed when
a) maleic acid and
b) fumaric acid
is treated with bromine in the cold and in the absence of light? Justify your answer from a consideration of the reaction mechanism.

❓ 63

Mark the stereogenic units in lumefantrine (an antimalarial drug) and state how many compounds can be represented by this formula. How many stereoisomers in principle are possible with this constitution?

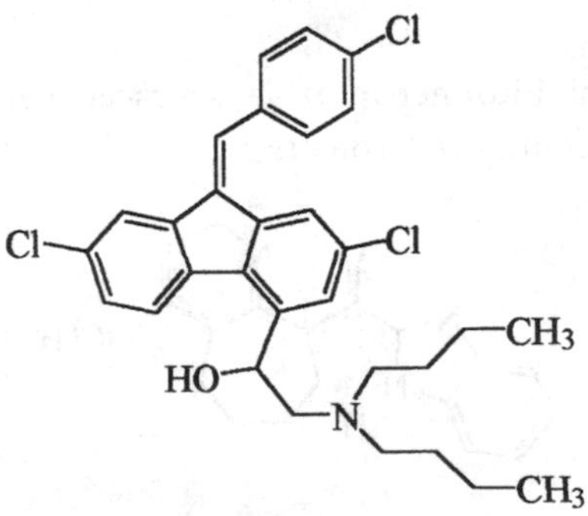

? 64

How many epimers of *trans*-1,2-dibromocyclopentane exist?

? 65

Draw as a Newman projection (*R*)-2-methylbutane-1-thiol with an antiperiplanar conformation along the C1-C2 bond.

? 66

Draw the structure of *meso*-tartaric acid (2,3-dihydroxybutanedioic acid) with an anticlinal conformation in a sawhorse projection.

? 67

How many stereoisomers are there of 4-*sec*-butylcyclohexanol? Give a reason.

? 68

Determine all the stereoisomers of the coordination entity shown below and assign appropriate stereodescriptors.

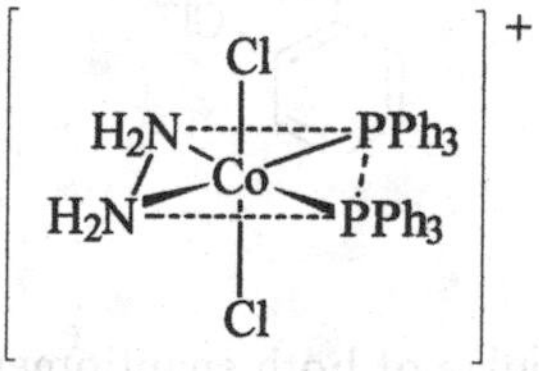

? 69

Determine the conformation of the following compound and convert the representation into a Fischer projection formula.

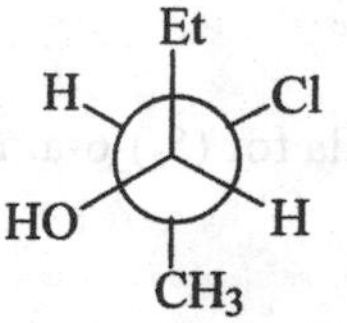

❓ 70

Draw the following biphenyl derivative in a projection which corresponds to it being viewed along an axis passing through atoms 4, 1, 1' and 4' and predict whether the compound is chiral.

❓ 71

Determine the priority order of the groups at position 4 of the 1,4-dihydro-pyridine ring in the calcium antagonist amlodipine and state whether the compound is chiral.

❓ 72

Draw the structural formulae of both enantiomers of 2t,3c-dichlorocyclo-hexan-1r-ol and deduce the configuration at all of the chirality centres.

❓ 73

To which symmetry point group does 2-methylhex-3-yne belong?

❓ 74

Draw an unambiguous formula for (S_a)-6-aminospiro[3.3]heptan-2-ol.

❓ 75

Draw the products from the reaction of *trans*-2-bromo-4-chlorocyclobutan-one with LiAlH₄ by attack from the *Re* and from the *Si* sides. Deduce the absolute configuration of the reaction products.

❓ 76

How many stereoisomers are there of bis(2-aminoethanethiolato-*N*,*S*)-nickel(II)? Give appropriate stereodescriptors for all the isomers.

❓ 77

Draw the structural formula for (2*R*,3*R*,4*R*)-3-chloro-4-isopropyl-2-methyl-cyclohexanone in its chair form with the lowest energy.

❓ 78

In the hydrogenation of (*E*)-pent-2-ene and (*Z*)-pent-2-ene, which iso-mer produces the larger exotherm? In the hydrogenation of (*E*)- and (*Z*)-cyclooctene, which isomer liberates the most heat?

❓ 79

Draw an unambiguous structural formula for the cytostatic (*SP*-4-3)-amminedichlorido(2-methylpyridine)platinum.

❓ 80

Draw the structural formulae for both (*RS*,*RS*)- and (*RS*,*SR*)-2-phenyl-2-(piperidin-2-yl)acetic acid methyl ester.

❓ 81

Determine the absolute configuration of the chirality centres in the anti-biotic tazobactam.

❓ 82

The reaction of (S)-1-methylheptyl tosylate with sodium azide yields a product with an enantiomeric excess (ee) of 99 %. What is the configuration of both the major and minor product? What is the probable cause for the small amount of contamination of the major product by the other enantiomer?

❓ 83

Specify the configuration of the stereogenic units in the anti-ulcer agent trimoprostil.

❓ 84

Determine the configuration of the gyrase inhibitor trovafloxacin. Is the compound chiral?

❓ 85

Draw the structural formulae of the products from the reaction of bicyclo[2.2.2]octene with
a) a peracid and
b) with potassium permanganate.
Show whether the products are chiral or achiral.

❓ 86

Deduce the symmetry point group of (S,S)-tartaric acid in the +synclinal conformation.

❓ 87

Assign the configuration of the stereogenic units in ataprost, an inhibitor of platelet aggregation.

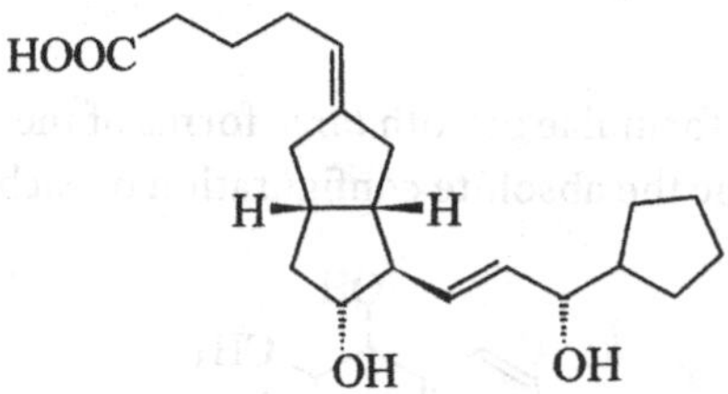

❓ 88

How many peaks would be expected in the chromatogram of the compound shown below when it is analysed on a chiral stationary phase? Give a reason.

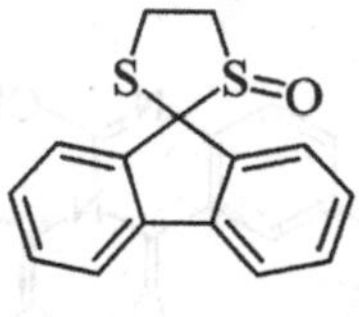

❓ 89

Determine the configuration of the chirality centres and double bonds in the vitamin D derivative maxacalcitol, and indicate whether the structural formula represents the s-cis or the s-trans isomer.

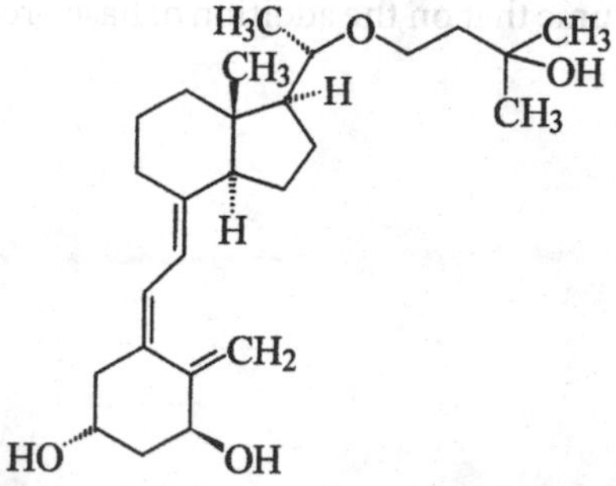

❓ 90

What is the product obtained from the reduction of (2S,3R)-2,3-dichloro-cyclobutanone with LiAlH₄ by attack from the *Re* side?

❓ 91

Draw the structural formula of (*R*)-2-bromopentan-3-one and describe the topicity of the methylene and methyl hydrogen atoms.

❓ 92

Draw the structural formulae of both *threo* forms of the appetite depressant cathin and determine the absolute configuration of each isomer.

❓ 93

Draw the structural formula of the R_a-configured muscle relaxant aflo-qualone.

❓ 94

Draw the structural formulae of the products from the aldol reaction of acetaldehyde and propiophenone and determine the configuration of the chirality centres. Assume that on the addition of base propiophenone attacks the acetaldehyde.

❓ 95

Label the hydrogen atoms at the prochirality centres in the following formulae with *pro-R* or *pro-S*.

a)

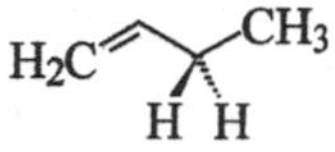

H_2C ═ CH_3
H H

b)

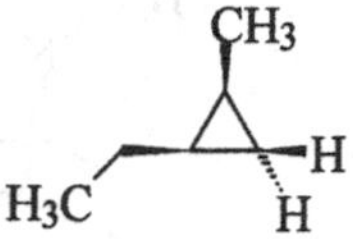

CH_3

H_3C ─── ───H
H

c)

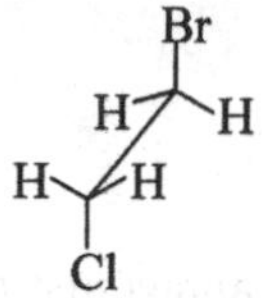

Br
H   H
H   H
Cl

d)

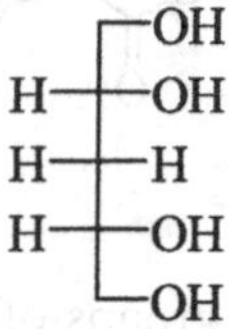

─OH
H──OH
H──H
H──OH
─OH

e)

$COOH$

H─ ─H

NH_2

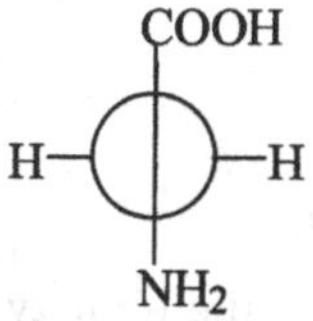

❓ 96

In the *N*-oxide of loperamide (an antidiarrheal) the hydroxy group and the oxide oxygen have a *trans* relationship. Draw the structure in a chair form and determine whether the molecule is chiral. Which stereodescriptors must be added to the systematic name to exactly describe the configuration?

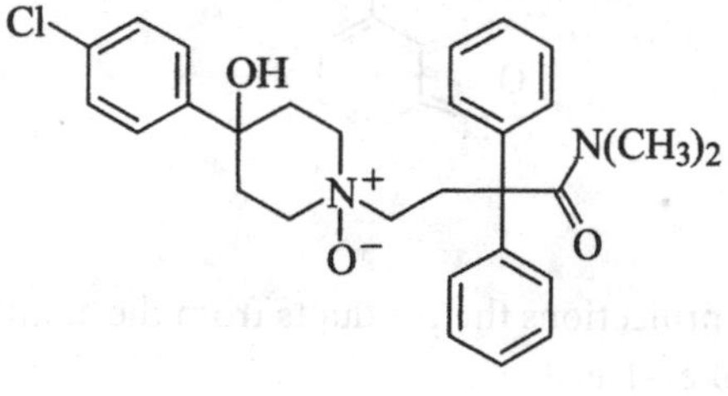

Cl

OH

N^+

O^-

$N(CH_3)_2$

O

❓ 97

What products are obtained from the bromination of cinnamic acid [(*E*)-3-phenylpropenoic acid]? Draw the structural formulae of the products as Fischer projection formulae. What relationship do the products have to one another?

❓ 98

Determine the configuration of the pseudochirality centre in the 5-HT$_3$ antagonist tropisetron whose structure is shown in the formula below.

❓ 99

Deduce the topicity of the methyl groups of the calcium antagonist darodipine by replacing one hydrogen atom in one of the methyl groups by a deuterium atom. Determine the relationship of the resulting compounds.

❓ 100

Draw as Haworth projections the products from the addition of bromine to (R)-4-chlorocyclohex-1-ene.

❓ 101

Determine the absolute configuration of the chirality centres in the anticonvulsant dizocilpine.

❷ 102

Different cyclic products are formed in the double Michael addition of malonic acid ethyl methyl ester to (*E,E*)-1,5-diphenylpenta-1,4-dien-3-one under basic conditions. Label the stereogenic units in the reaction products with the appropriate stereodescriptors.

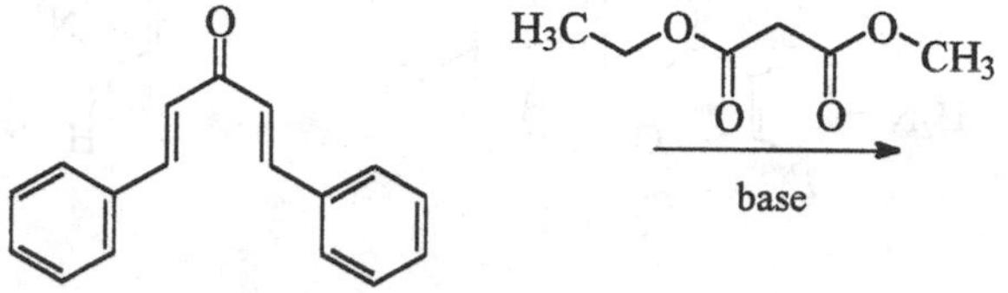

❷ 103

Determine the absolute configuration of the beta-lactam antibiotic sulopenem.

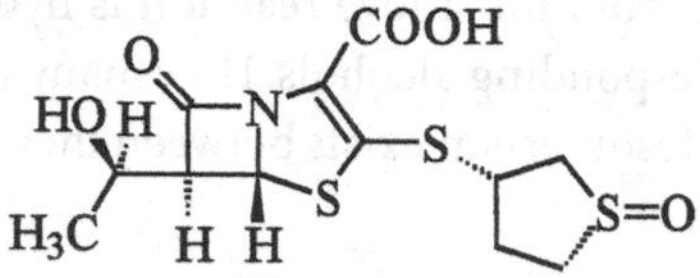

❷ 104

Are the two faces of the double bonds in the compounds represented by the following formulae homotopic, enantiotopic or diastereotopic? Give, where appropriate, a suitable descriptor for the face oriented towards you.

a)

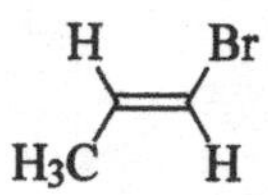

b)

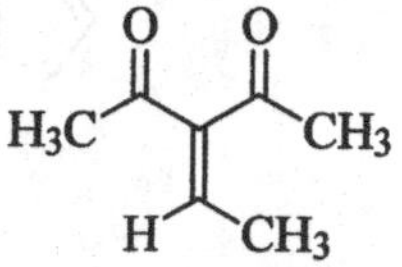

c)

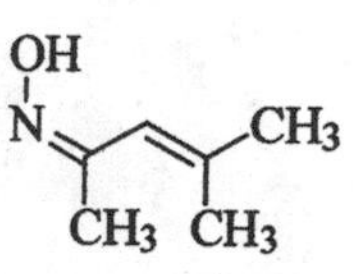

d)

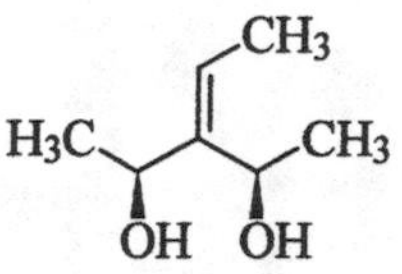

❓ 105

Which stereogenic units are present in the cephalosporin antibiotic cefmatilen? Determine their configuration.

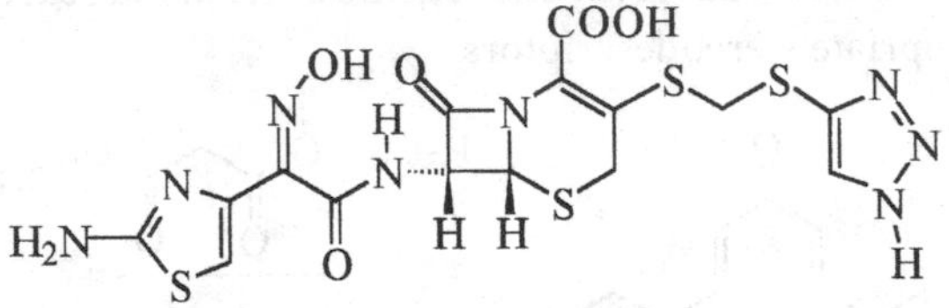

❓ 106

Maleic anhydride is subjected to a cycloaddition reaction with cyclopenta-
1,3-diene. What products would be produced and which stereodescriptors
could be used to completely describe these compounds? Are the products
chiral?
The material obtained from the above reaction is hydrolysed and then reduced to yield the corresponding alcohols. How many isomeric products are
obtained? What type of isomerism exists between them and how could these
products be separated?

❓ 107

Draw the structural formula of the R_a-configured atropisomer of the compound represented by the formula below.

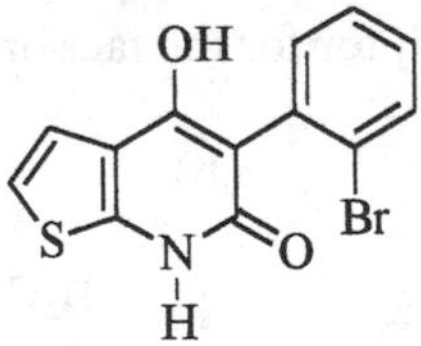

❓ 108

Assign suitable stereodescriptors to the dopamine reuptake inhibitor brasofensine.

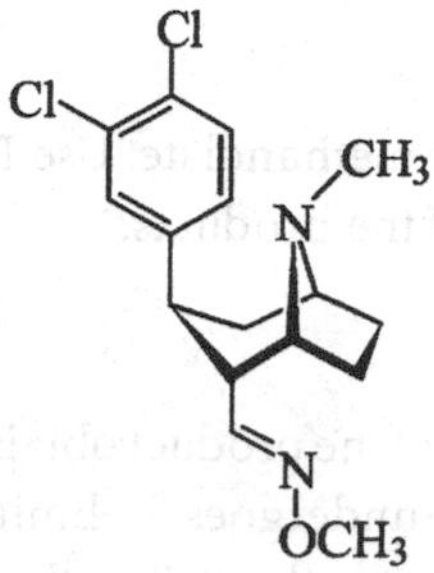

❓ 109

Draw (2R,3s,4S)-2,3,4-trichloropentanedioic acid as a Fischer projection formula.

❓ 110

Determine the configuration of the chirality centres in the HMG-CoA reductase inhibitor lovastatin.

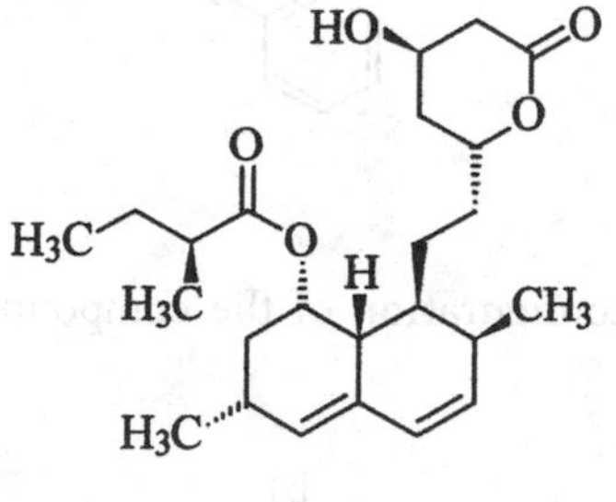

❓ 111

Which stereogenic unit is present in the compound the formula of which is shown below and which configuration can be assigned?

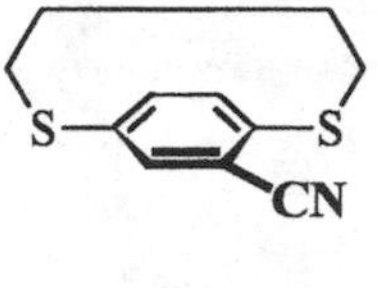

❓ 112

Draw the structural formulae of the products which are obtained when the bromonium ion obtained from
a) maleic acid and
b) fumaric acid,
and bromine is treated with methanolate. Use Fischer projection formulae to compare the structures of the products.

❓ 113

Draw the structural formula of the product obtained when (1S,2R)-1-bromo-2-fluoro-1,2-diphenylethane undergoes β-elimination of HBr. What product is obtained when the substrate is the R,R or the S,S isomer?

❓ 114

Determine the configuration of the stereogenic units in the depicted form of the 5-HT₃ antagonist eplivanserin.

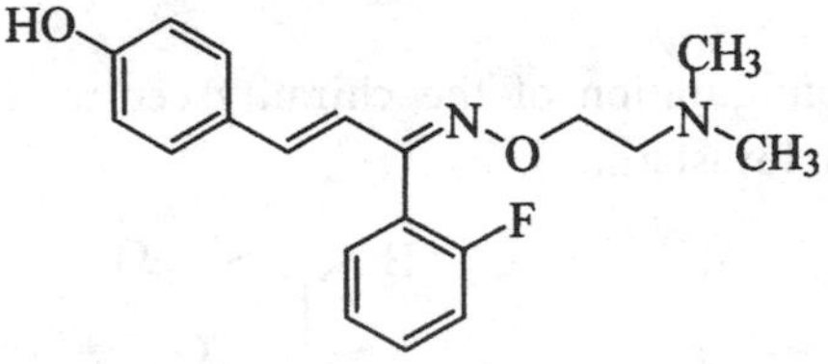

❓ 115

Deduce the absolute configuration of the compounds represented by the following formulae.

a)

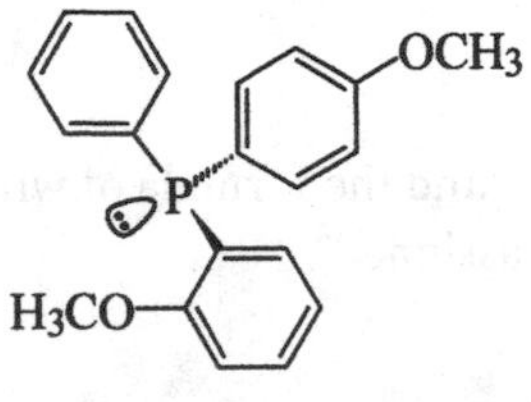

b)

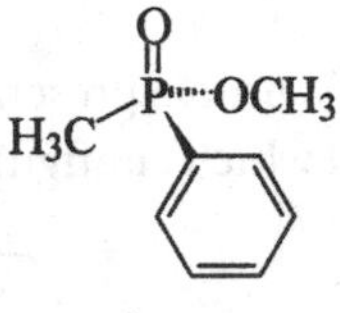

c)

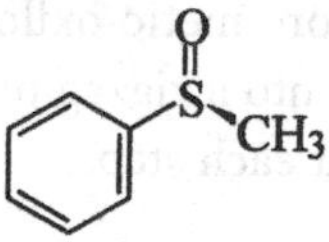

d)

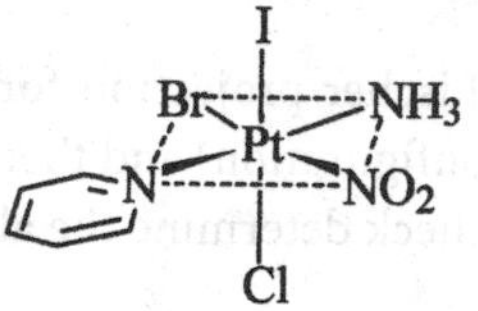

❓ 116

The naturally occurring substance besigomsin has an R_a-configured chirality axis. Draw the structural formula of this isomer.

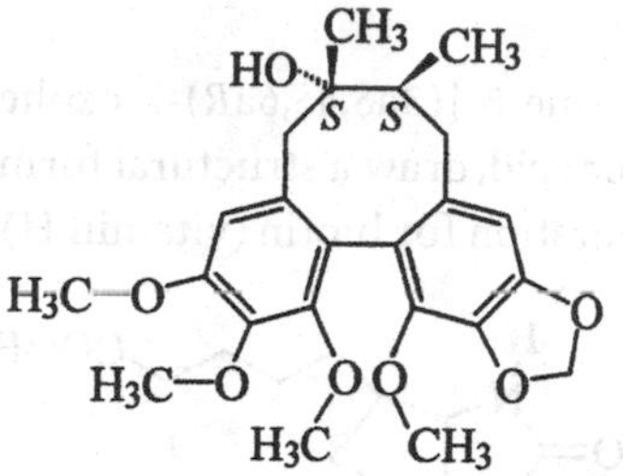

❓ 117

Latanoprost is a prodrug used in the treatment of glaucoma. The 15S isomer of the acid is only about 10 % as active as the 15R isomer of the acid. Draw the structural formula of the more active epimer and determine the configuration of the remaining stereogenic units present.

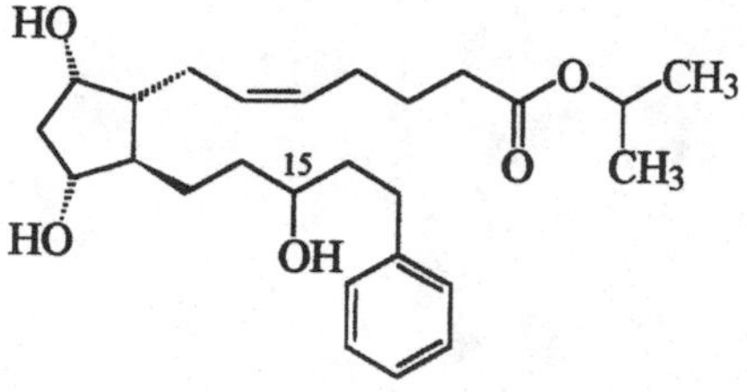

❓ 118

Draw the Fischer projection formula for the sympathomimetic oxilofrine (*erythro* configuration) and then convert this formula into a zigzag projection. As a check determine the absolute configuration at each step.

❓ 119

Given its systematic name 5-[(3a*S*,4*S*,6a*R*)-2-oxohexahydrothieno[3,4-*d*]-imidazol-4-yl]pentanoic acid, draw a structural formula with complete specification of the configuration for biotin (vitamin H).

❓ 120

In the formula shown below for rodorubicin (a cytostatic agent) convert the groups shown in the Haworth projection into a planar projection of the rings in which the substituents are shown with wedged bonds.

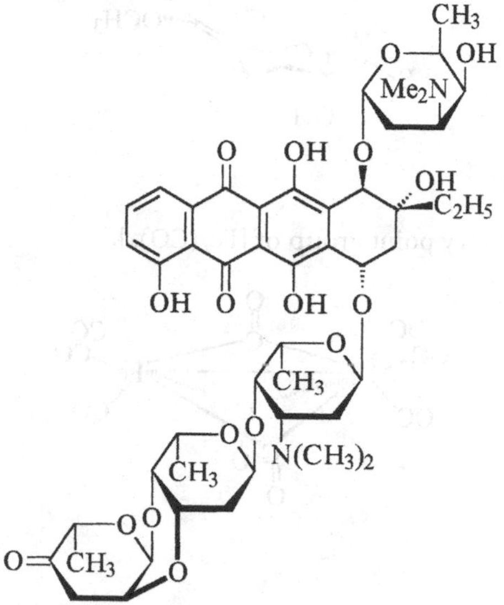

❓ 121

From the constitutional formulae given below, first of all draw formulae for all the possible configurational isomers and then deduce how many sets of chemically equivalent hydrogen atoms are present in the represented compounds.

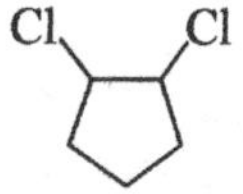

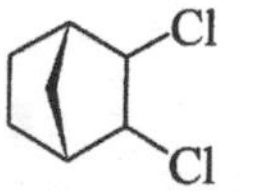

❓ 122

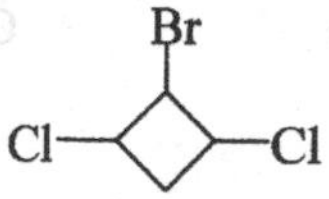

What products are obtained from the reaction between benzaldehyde and butanone in the presence of base (NaH, room temperature)? Draw the structural formulae of all of the reaction products in spatially correct Newman projections. Chose the conformation in which the phenyl and carbonyl groups are antiperiplanar to one another.

❓ 123

Which stereogenic unit is present in 1-(bromomethyl)-4-[chloro(methoxy)-methylidene]cyclohexanol? Determine the absolute configuration of the isomer shown below.

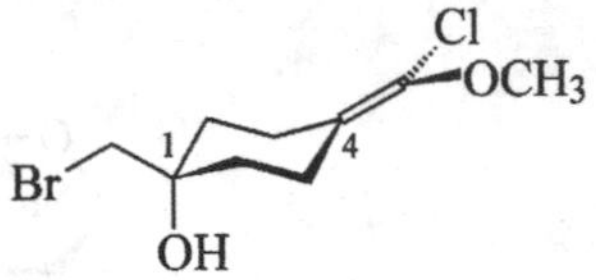

❓ 124

Deduce the symmetry point group of $[Fe_2(CO)_9]$.

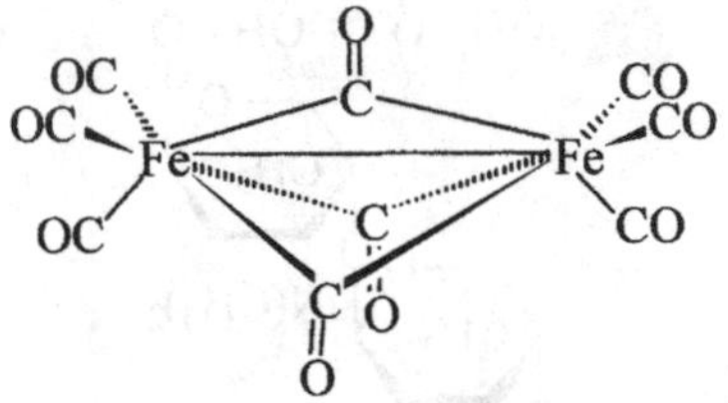

❓ 125

Describe the base induced elimination reactions of the deuterated compounds represented by the following formulae and assign appropriate stereo-descriptors to the resulting olefins. (Note that as well as the hydrogen atom the deuterium atom can also be abstracted from these compounds.)

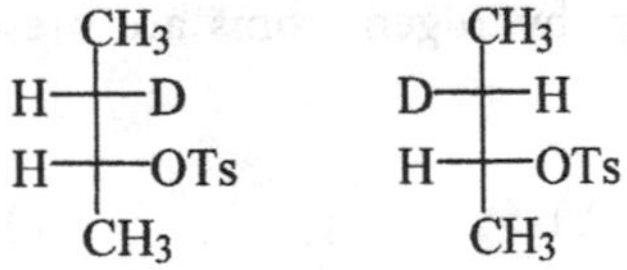

❓ 126

Draw the formulae (sawhorse projection) of all the chiral conformations of 2-chloroethanol and name them.

❓ 127

Specify the configuration of the chirality centres in the antimalarial drug cinchonine.

❓ 128

The room temperature ^{1}H NMR spectrum of thiophene-3-carboxamide contains five signals. Give a reason for this observation.

❓ 129

What products are obtained when 1-methylcyclopenta-1,3-diene reacts as a diene with the dienophile maleic anhydride?

❓ 130

(1RS,2RS,4RS)-1,7,7-Trimethylbicyclo[2.2.1]heptan-2-yl acetate is a hyperaemic agent. Draw the structural formula(e).

❓ 131

Draw the structural formula of cis-1-[(R)-sec-butyl]-2-methylcyclohexane in its lowest energy chair conformation. How many isomers of this compound are there?

❓ 132

Which stereogenic unit is present in the following adamantane derivative? Determine its configuration.

? 133

Draw the structural formula of (*SP*-4-1)-bis(glycinato-*N*,*O*)platinum(II). (Glycinate is the anion of the amino acid glycine [aminoacetic acid].)

? 134

(*S*)-Pentan-2-amine is required for a reaction. However, only the enantiopure alkanols, (*R*)- and (*S*)-pentan-2-ol, are available. How could the desired amine be prepared using either one of these?

? 135

Draw the antiperiplanar conformation of *erythro*-3-chloro-2-hydroxybutanoic acid in Newman projection.

? 136

Draw the structural formula of 2,2-dichloro-1,1-difluoro-4,4-dimethylcyclohexane as a Newman projection formula viewed along the C1-C2 and C5-C4 bonds. Assume that the compound adopts a chair conformation.

? 137

Which are the epimers of (2*R*,4a*R*,8a*R*)-decahydronaphthalen-2-ol? Draw their structural formulae in the chair conformation.

? 138

Show that the ethylenediaminetetraacetate complex of Ca^{2+} belongs to the symmetry point group C_2 and assign a suitable stereodescriptor to the compound.

? 139

Which configuration has *sn*-glycerol-3-phosphate in the *R*/*S* system?

? 140

The vasopeptidase inhibitor omapatrilat is a pure stereoisomer and its systematic name is (4*S*,7*S*,10a*S*)-5-oxo-4-{[(2*S*)-3-phenyl-2-sulfanylpropanoyl]amino}octahydropyrido[2,1-*b*][1,3]thiazepine-7-carboxylic acid. Draw the structural formula of this compound with complete specification of the configuration. Which amino acid residue is present in the molecular structure of this compound?

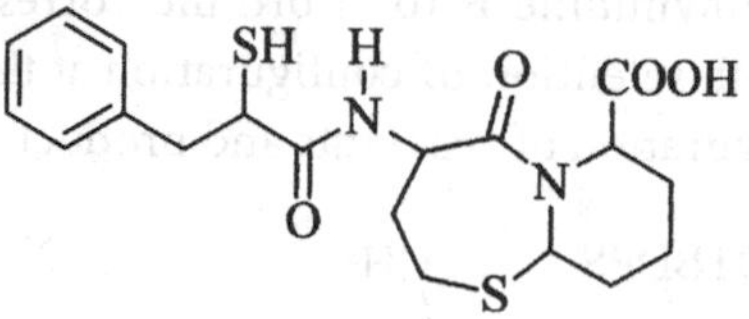

❓ 141

Draw the structure of (R,R)-3-chloro-4-fluoro-1,1-dimethylcyclohexane as a Newman projection formula viewed along the C6-C1 and C4-C3 bonds. (The chlorine and fluorine atoms should be drawn antiperiplanar to each other.)

❓ 142

Give an unambiguous stereodescriptor for the anion of the compound shown below.

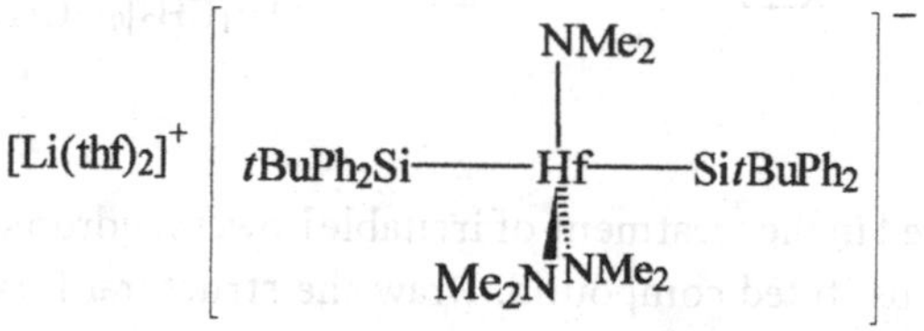

❓ 143

Deduce the configuration of the molecule represented by the formula below.

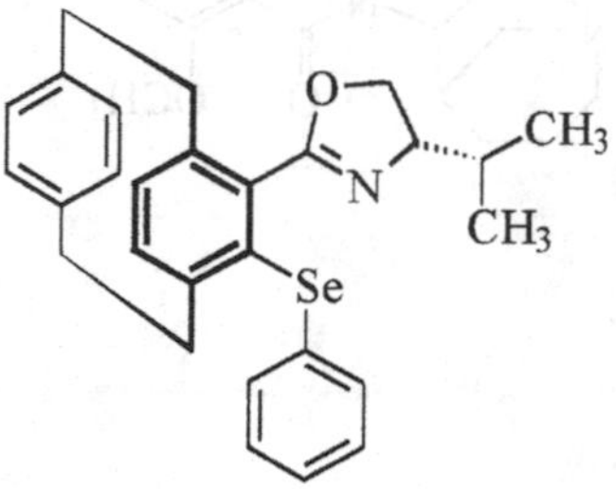

? 144

It has been shown that coupling of the ethyl phosphinate **A** with 5'-azido-2'-methoxy-5'-deoxythymidine **B** to afford the corresponding phosphonamidate proceeds with retention of configuration at the phosphorus atom. Determine the configuration of reactants and product.

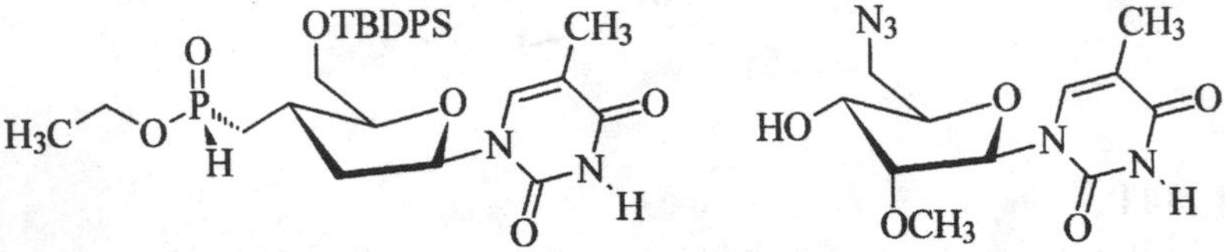

A   **B**

? 145

What is the configuration of the compound represented by the following formula?

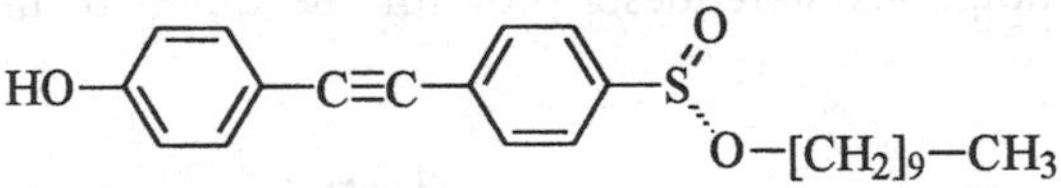

? 146

Renzapride used in the treatment of irritable bowel syndrome is a racemate of the *endo* substituted compound. Draw the structural formulae of both isomers.

❷ 147

Deduce whether enniatin B (a compound with antiretroviral activity) is chiral and determine its symmetry point group.

❷ 148

Which diastereomer of 1,3-dichlorocyclopentane has a ^1H NMR spectrum which contains four signals with relative intensities 1:1:1:1?

❷ 149

Deduce whether the appetite depressant levofacetoperane has the *threo* or *erythro* configuration.

❓ 150

Determine the configuration of the stereogenic units in rifaximin, a rifamycin antibiotic used in encephalopathy.

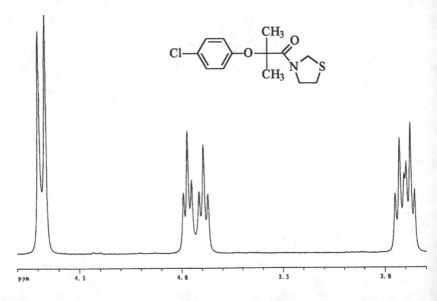

❓ 151

Explain why the ¹H NMR spectrum of the 1,3-thiazolidine derivative of clofibric acid (300 MHz, CDCl₃, room temperature) contains, in addition to two doublets for the aromatic protons and a sharp singlet at 1.61 ppm for the methyl groups, the complex signal splitting pattern (shown below) in the region between 2 and 5 ppm.

❓ 152

What products would be expected from the reaction of the compound the formula of which is shown below with lithium [dimethyl(phenyl)silyl]-iodidocuprate(I)? Determine their configuration.

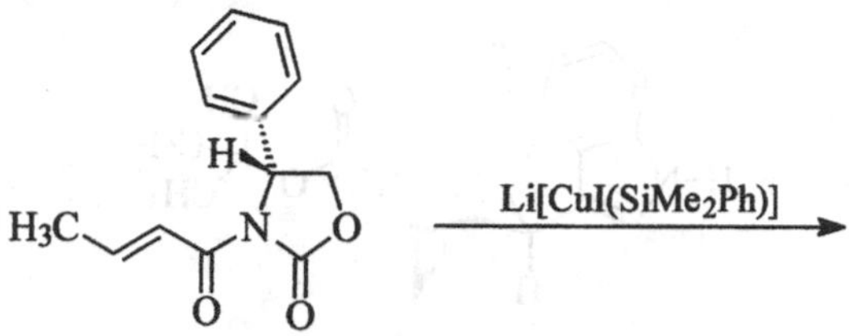

$$\text{Li[CuI(SiMe}_2\text{Ph)]}$$

❓ 153

Determine the configuration of the chirality centres in the ACE inhibitor fosinopril.

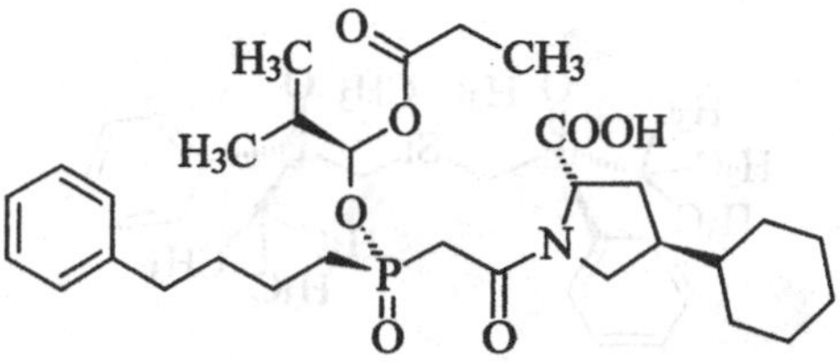

❓ 154

Draw formulae for all the isomers of dichlorocyclopropane and predict the relative signal intensities in their ^{1}H NMR spectra. Compare your answer with results obtained from a determination of the symmetry elements and point groups.

❓ 155

Identify the stereogenic units in the molecules represented by the following formulae, assign appropriate stereodescriptors and indicate what type of isomerism is possible in these compounds.

a)

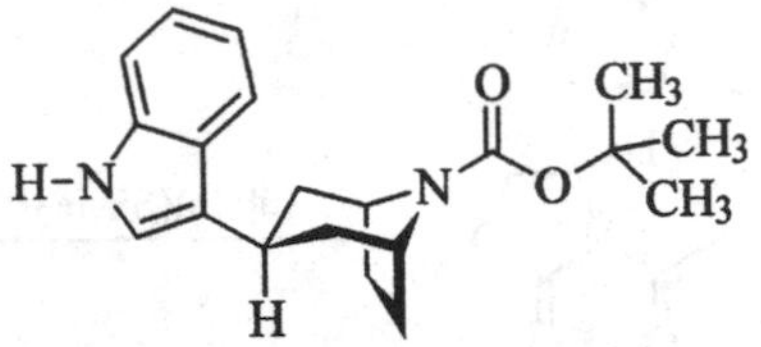

b)

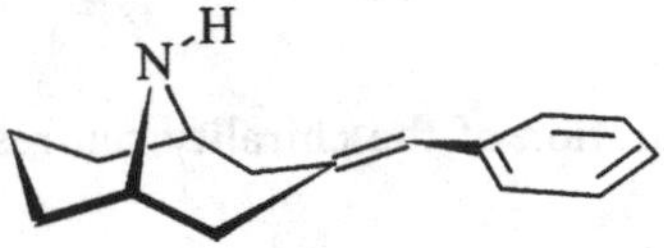

c)

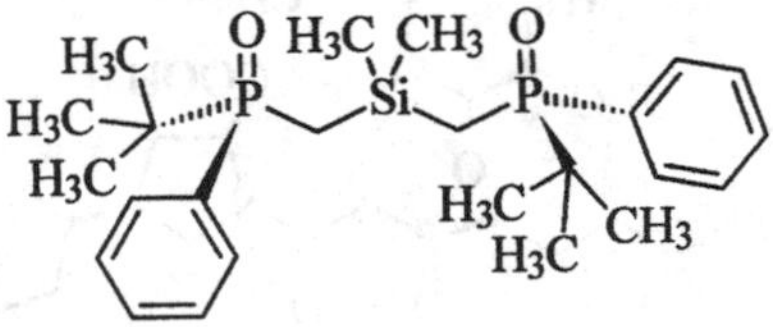

d)

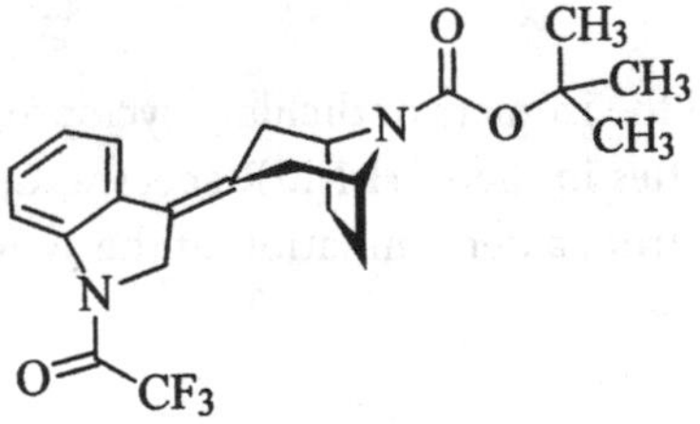

e)

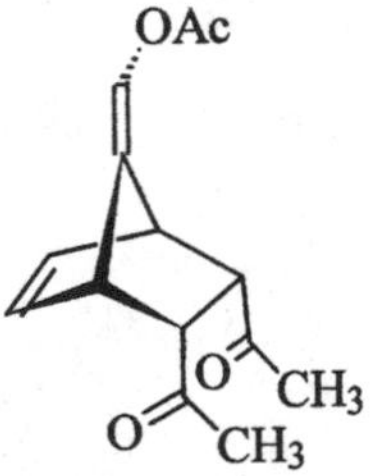

❓ 156

Determine the configuration of the chirality centres in the compound represented by the formula shown below. To which symmetry point group does the compound belong? Use the flow chart in the appendix to assist you.

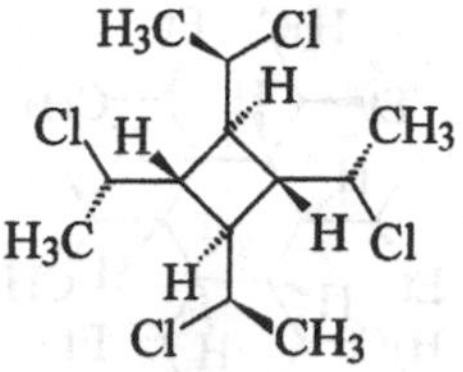

❓ 157

Describe the configuration of the molecule shown below by means of the appropriate stereodescriptors.

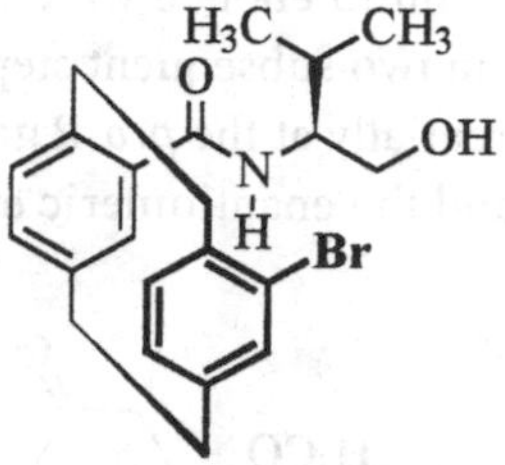

❓ 158

Draw the structural formula of 2-methyl-2-(3-oxobutyl)cyclopentane-1,3-dione and determine the prochirality centres and the topicities of the hydrogen atoms.

❓ 159

Cyanohydrins were obtained from the reaction of racemic 2-methylbutanal with hydrogen cyanide. What products are obtained preferentially?

❓ 160

Determine the configuration of the chirality centres in the compound shown below. To which symmetry point group does it belong? Use the flow chart in the appendix to assist you.

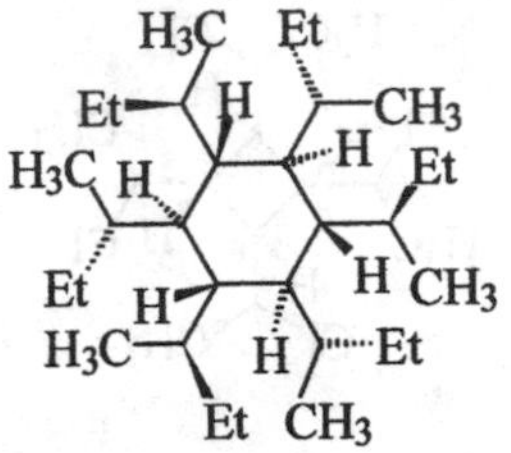

❓ 161

By means of a chiral base the compound shown below can be converted enantioselectively into its lithium enolate which can be transformed into an α,β-unsaturated ketone in two subsequent steps. If deprotonation of the initial ketone occurs preferentially at the *pro-R* group to the extent of 92 %, what is the configuration and the enantiomeric excess of the resulting α,β-unsaturated ketone?

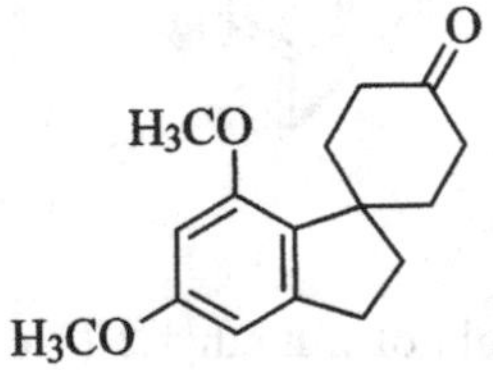

❓ 162

Draw the structural formula of (S)-2-methoxytetrahydropyran in its most energetically favourable chair form and give the reason for your answer.

? 163

How many stereoisomers result from the non-selective reduction of the following substituted cyclobutanedione with $LiAlH_4$? What are the topicities of the hydrogen atoms at positions 1 and 3 in the reaction products?

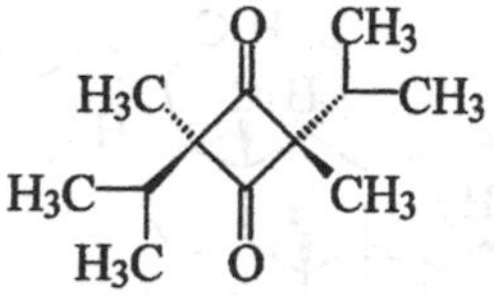

? 164

Determine the symmetry point group of $[Co_4(CO)_{12}]$.

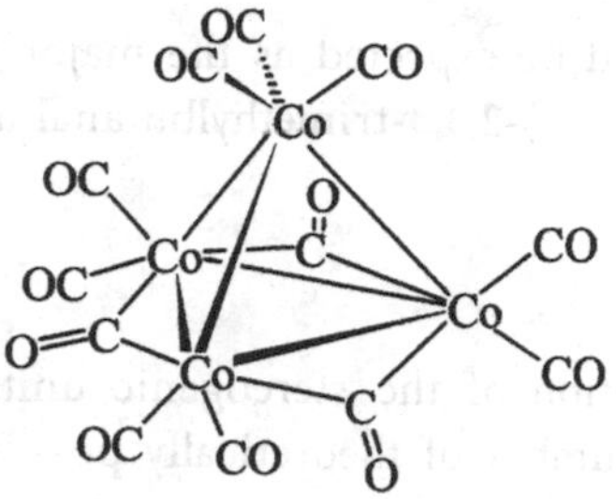

? 165

How many products would be expected from the reaction between (2E,4Z)-hexa-2,4-diene and 2-methoxycyclohexa-2,5-diene-1,4-dione? What isomeric relationships do the resulting products have to one another?

? 166

Determine the absolute configuration and the number of possible isomers of the compound shown below.

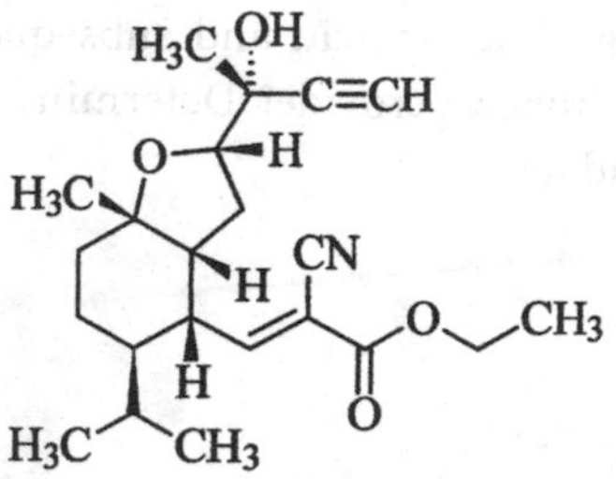

❓ 167

Does the methoxy group in the compound shown below occupy an axial or an equatorial position? Use the configuration of the compound to deduce your answer.

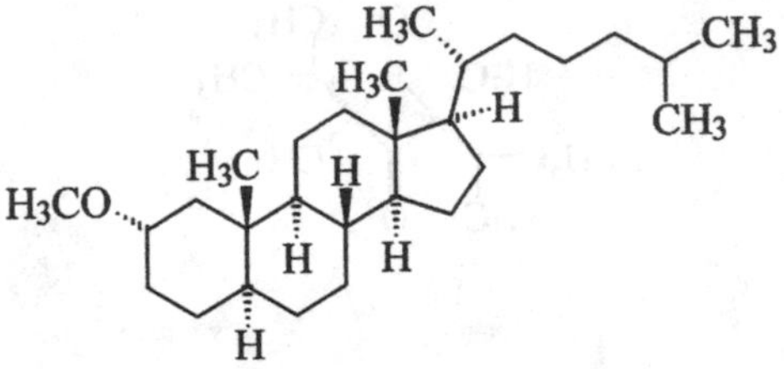

❓ 168

What compound would be expected as the major product from a Grignard reaction between (S)-2,3,3-trimethylbutanal and propylmagnesium bromide?

❓ 169

Deduce the configuration of the stereogenic units of the antipsoriatic calcipotriol and the number of theoretically possible stereoisomers with this constitution.

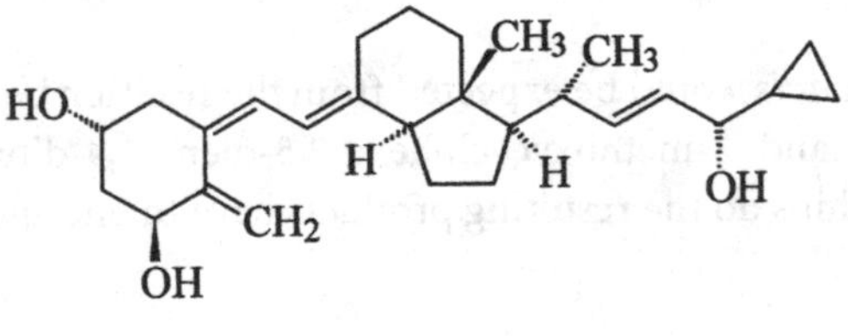

❓ 170

What product is obtained from the reaction of 1-methylcyclopentene with $NaBH_4$ in the presence of acetic acid and subsequent oxidation with an alkaline solution of hydrogen peroxide? Determine also the configuration of the intermediate product.

❓ 171

Identify the stereogenic units of the molecules represented by the following formulae and assign appropriate stereodescriptors.

a)

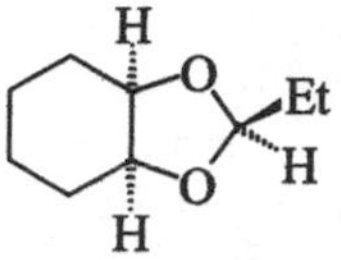

b)

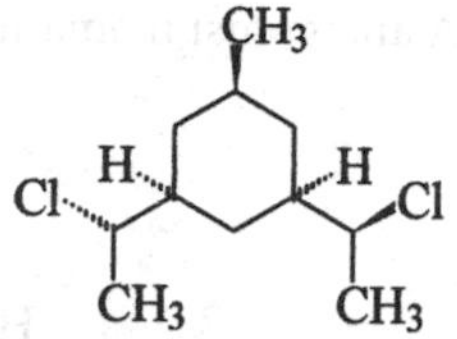

c)

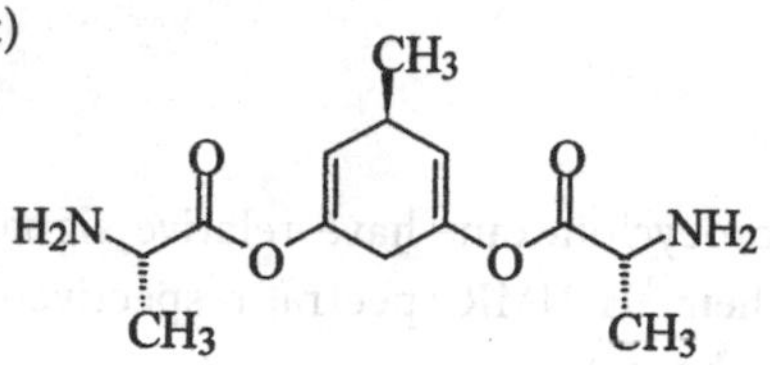

d)

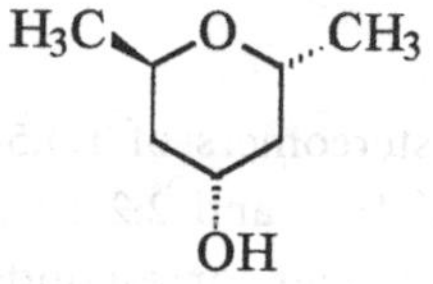

e)

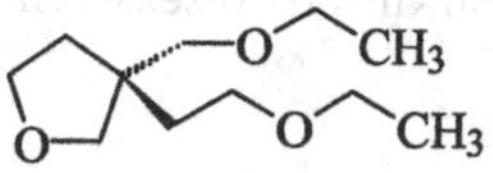

f)

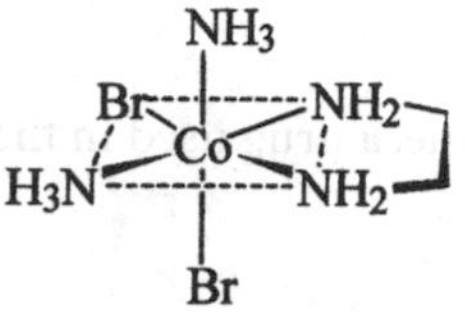

❓ 172

The reaction of the tetrabutylammonium salt of cAMP with 4-(bromo-methyl)-2H-chromen-2-one yields two products which result from axial and equatorial attack of the phosphate group. What is the isomeric relationship between the two products? Determine the configuration of all the chirality centres in both products.

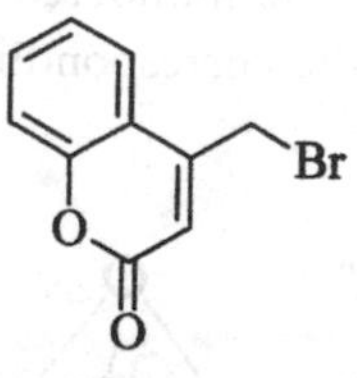

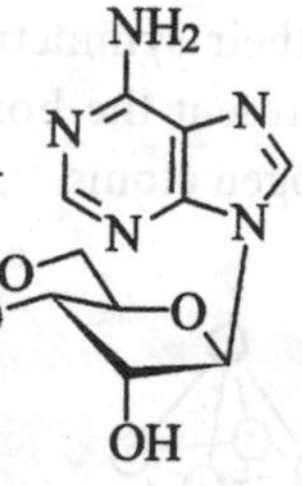

cAMP

❓ 173

Which symmetry elements are present in *meso*-tartaric acid?

❓ 174

Is the NMDA antagonist memantine chiral?

❓ 175

The two diastereomers of 1,3,5-trichlorocyclohexane have relative signal intensities of 1:1:1 and 2:2:2:1:1:1 in their ^{1}H NMR spectra, respectively. Which set of signals corresponds to which isomer?

❓ 176

Is tropatepine, a drug used in the treatment of Parkinson's disease, chiral?

❓ 177

Determine the symmetry elements present in the following boranes and hence assign their symmetry point groups. The numbered circles in the polyhedra represent the boron atoms with the corresponding number of attached hydrogen atoms.

a) B_5H_9

b) B_4H_{10}

c) B_6H_{10}

d) B_5H_{11}

❓ 178

Kinetic data indicate that the alkaline hydrolysis of the compound represented by the formula shown below proceeds by an S_N2 mechanism. What is the configuration of the starting material and of the resulting sulfoximine?

❓ 179

Deduce the symmetry point group of $[(PdCl_2)_6]$.

❓ 180

Determine the absolute configuration of the chirality centres in the antibiotic doxycycline.

❓ 181

Draw the structural formulae of both pyranose forms which would result from ring closure (hemiacetal formation) of D-idose. What is the relationship between these structures and the corresponding pyranose forms derived from L-idose? Check your answer by determining the configuration of the chirality centres according to the R/S nomenclature.

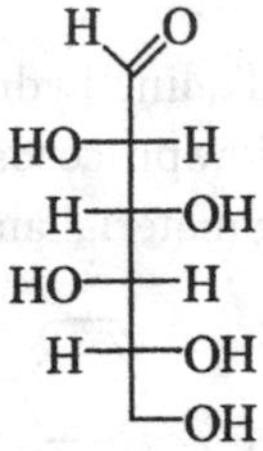

❓ 182

Show that $[CoCl_2(en)_2]$ (en = ethane-1,2-diamine) can be racemic. Determine the symmetry point groups of each isomer and assign suitable stereo-descriptors.

❓ 183

Which pyranoses react with phenylhydrazine to yield the osazone the formula of which is shown below?

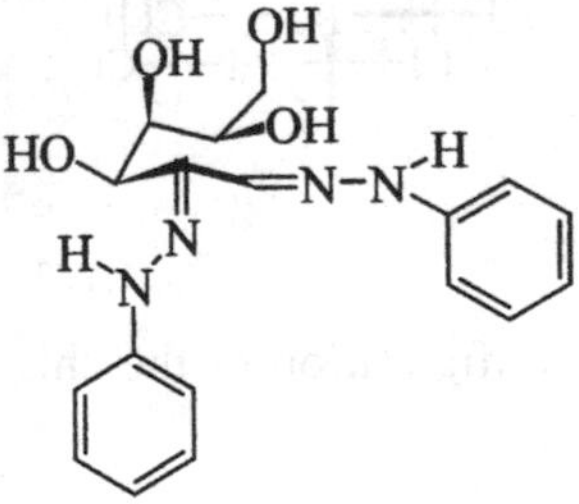

❓ 184

The muscle relaxant mivacurium chloride is a mixture of stereoisomers. Calculate how many isomers are theoretically possible with the constitution shown below.

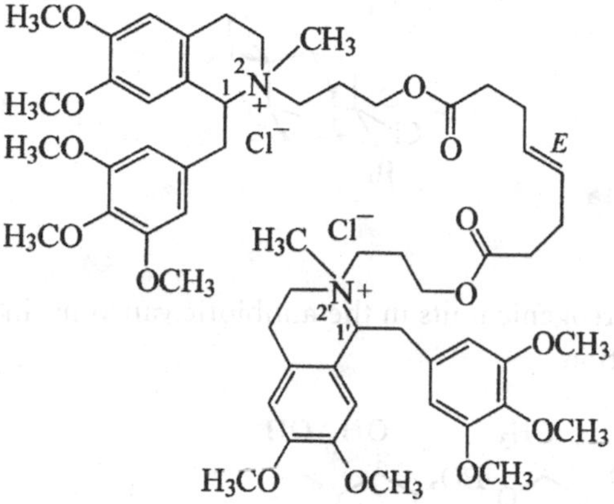

❓ 185

Deduce the symmetry point group of the following idealised representation of the structure of copper(I) benzoate. (Assume that the plane of the phenyl rings lies parallel to the carboxylate groups.)

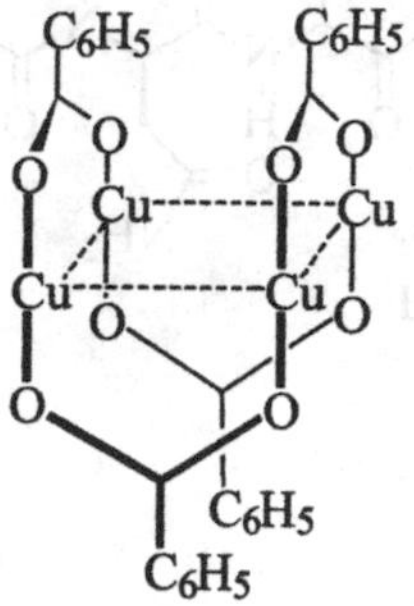

❓ 186

Determine unequivocally the configuration of 1,6-dibromo-3,6-dichloro-adamantane. How many stereoisomers of this compound exist and to which symmetry point groups do they belong?

❓ 187

Identify the stereogenic units in the antibiotic vancomycin and determine their configuration.

188

Compound A can be converted into compound B in five steps by standard reactions. Show all the steps in the reaction sequence, deduce the absolute configuration of all the intermediate compounds and indicate whether the reactions proceed with retention or inversion of configuration.

A

B

189

Devise a synthesis for the pure enantiomers of (R)- and (S)-methyloxirane.

190

Devise a synthesis for the monoamine oxidase inhibitor tranylcypromine (rac-A).

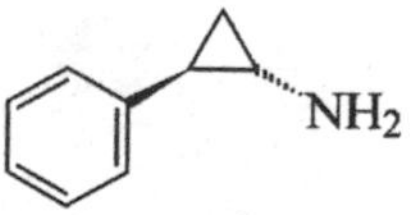

A

191

The β-receptor blocker propranolol is still used as a racemate, although the S enantiomer is the active compound. Devise a synthesis for this enantiomer.

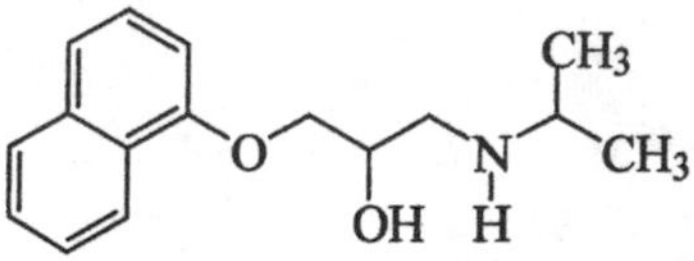

188

Compound A can be synthesized from compound B in five steps by standard reactions. Show all the steps in the reaction sequence, deduce the absolute configuration of all the intermediate compounds and indicate whether the reaction proceed with retention or inversion of configuration.

189

Devise a synthesis for the four enantiomers of (R)- and (S)-isethionolactone.

190

Devise a synthesis for the monomeric aldose inhibitor drug from compound A.

191

The precursor acid erythptan, is esterified with acetone, although the reaction is a bit sluggish, and Dean-Stark conditions with a hemiacetal.

Answers

❶ 1

a) **Chirality** is the property of an object (e.g. a molecule) of being non-superimposable on a mirror image like object.

b) The **constitution** of a compound is the number and types of atoms contained in the compound along with information about how the atoms are connected to each other, and their bond order. It does not convey any information regarding their spatial arrangement.

c) The **configuration** of a molecule is the spatial arrangement of atoms or groups of atoms in the molecule and is independent of rotation about any single bond.

d) The **conformation** of a molecule is the precise spatial arrangement of the atoms or groups of atoms in a molecule as a result of rotation about single bonds. There are an infinite number of possible conformations. However, only those conformational isomers possessing energy minima are referred to as **conformers**.

e) **Stereoisomers** are isomers with the same constitution but with different spatial arrangements of their atoms or groups. Stereoisomers may be subdivided into configurational isomers and conformational isomers. In both types the isomers are either enantiomers or diastereomers.

2

There are nine constitutional isomers with the empirical formula C_3H_6O. However, since two of the compounds exist as pairs of stereoisomers, the total number of isomers equals eleven. The two methyloxiranes, **A** and **B**, are enantiomers, they are stable and very volatile. Acetone (**C**) is in equilibrium with its enol tautomer (propen-2-ol, **D**). Similarly, the two diastereomeric enols (*E*)- and (*Z*)-prop-1-en-1-ol (**E** and **F**) are in equilibrium with their tautomer propanal (or propionaldehyde, **G**). The other isomers are (last group of compounds from left to right): methoxyethene, prop-2-en-1-ol (commonly known as allyl alcohol), oxetane and cyclopropanol.

3

The **absolute configuration** is the actual spatial arrangement of the atoms or groups at a stereogenic unit of a chiral compound or substructure and is unambiguously described by means of appropriate **stereodescriptors**. The stereodescriptors used to denote the absolute configuration depend upon both the structure and the type of compound. In general, for tetrahedrally or trigonal pyramidally coordinated chirality centres, the two possibilities are described by the stereodescriptors *R* and *S*, whilst for compounds with other coordination geometry the stereodescriptors *A* and *C* are employed. For compounds possessing chirality axes or chirality planes, the stereodescriptors R_a and S_a, and R_p and S_p are used, respectively. Helical chirality is denoted by the descriptors *M* and *P*, and in the case of coordination compounds by Δ and Λ. The absolute configuration of amino acids and carbohydrates is invariably described by the small capital letters D and L.

For steroid ring systems and some other classes of natural products the non-italicised stereodescriptors α and β are most frequently encountered.

❶ 4

By application of the CIP rules the order of priority of the atoms directly attached to the chirality centre is O > C(O,O,(O)) > C(C,H,H) > H. The atom or group of lowest priority, hydrogen in this case, is already oriented away from the observer. Therefore the sequence of the remaining three groups can be directly deduced from the formula, and these are easily seen to be arranged in a counter-clockwise sense to the observer. It therefore follows that the formula represents (S)-2-hydroxysuccinic acid (formerly known as L-malic acid). The compound is produced in the citric acid cycle from fumaric acid by fumarate-hydratase (fumarase).

$$
\underset{\underset{1\qquad4}{\text{HO}\quad\text{H}}}{\text{HOOC}\overset{3\qquad2}{\diagup}\overset{S}{\underset{|}{\diagup}}\text{COOH}}
$$

❶ 5

a) **Enantiomerism** is the phenomenon that two non-superimposable objects (e.g. molecules) are related like mirror images. Two such non-superimposable mirror image like molecules are referred to as **enantiomers** (or mirror image isomers). Examples include (+)- and (−)-tartaric acid, and D- and L-alanine. For the determination of the configuration of one of the isomers of alanine see question 7.

b) **Diastereomers** are isomers which are constitutionally the same but are not related like mirror images (i.e. they are not enantiomers). Examples include fumaric acid and maleic acid, glucose and mannose, and (OC-6-21)- and (OC-6-22)-triamminetrinitrocobalt(III).

OC-6-21   OC-6-22

c) A **racemate** is an equimolar, and consequently an optically inactive, mixture of enantiomers. Examples include (*RS*)-methyloxirane, *rac*-1-phenylethanol, DL-alanine, the pharmaceutical debropol (see question 25) and racemic acid [a 1:1 mixture of (+)- and (−)-tartaric acids].

d) **Epimers** are diastereomers which have the opposite absolute configuration at just one chirality centre. One example is the inhibitor of platelet aggregation iloprost which is used clinically as a mixture of both epimers.

6

(*R*)-Carvone and (*S*)-carvone are enantiomers and can be distinguished from each other by their optical rotations, by circular dichroism and by smell. Laevorotatory (*R*)-carvone has a spearmint smell (spearmint = *Mentha spicata*), whilst (*S*)-(+)-carvone has a caraway odor.

7

The order of priority of the atoms and groups attached to the chirality centre by application of the CIP rules is N > C(O,O,(O)) > C(H,H,H) > H. The isomer depicted is therefore the unnatural amino acid (*R*)-alanine, more usually referred to as D-alanine.

❶ 8

a) **Symmetry elements** are sub-groups of the symmetry group of an object (e.g. a molecule). Every symmetry element comprises a certain type and number of symmetry operations which converts the object into a congruent (superimposable) arrangement. The internal symmetry elements which occur in a molecule are proper axes of symmetry (rotation axes) as well as symmetry elements of the second kind, i.e., planes of symmetry (mirror planes), centres of symmetry (inversion centres) and alternating axes of symmetry (rotation-reflection axes). The total of the symmetry elements of a molecule is its **point group**.

b) A **meso compound** is an achiral diastereomer in a group of stereoisomers which also contains chiral isomers. It contains at least one symmetry element of the second kind (often a plane of symmetry) which transforms enantiomorphic parts of the molecule into each other.

❶ 9

Lovastatin has eight chirality centres. The configuration of this compound is the subject of question 110.

❶ 10

a) **Mutarotation** is the change in optical rotation of a compound with time at a constant wavelength of plane polarised light. The cause of this phenomenon is a partial or complete change of one chiral compound into another compound of different configuration and/or constitution.

b) A reaction is said to be **enantioselective** if one of the two possible enantiomers is formed preferentially or exclusively.

c) The term **retention** indicates that the spatial arrangement of atoms or groups at a chirality centre remains unaltered relative to a reference group. For example, in a nucleophilic substitution reaction the group undergoing substitution serves as the reference group.

d) A **stereogenic unit** is a structural unit in a molecule responsible for the occurrence of stereoisomers. The most frequently encountered stereogenic units are chirality centres and chirality axes. In addition, there are chirality planes and pseudochirality centres. Double bonds and rings with suitably arranged prochirality centres can also be stereogenic units.

❗ 11

A molecule is **chiral** when it cannot be superimposed on the mirror image like molecule. The prerequisite is that the molecule contains no symmetry elements of the second kind.

❗ 12

a) **Atropisomers** are conformers which, due to restricted rotation around a single bond, are capable of being separated. They possess a chirality axis the configuration of which is denoted by the stereodescriptors R_a (less frequently M) or S_a (less frequently P).

b) **Anomers** are epimers (diastereomers) which differ in absolute configuration at the **anomeric centre**, i.e., the chirality centre produced from the carbonyl carbon atom during the formation of a cyclic hemiacetal of a carbohydrate or an analogous compound. The configuration of the anomeric centre is denoted by the non-italicised stereodescriptors α or β relative to the **anomeric reference atom**. For simple cases (trioses to hexoses) the anomeric reference atom is the same as the configurational reference atom which determines whether the sugar belongs to the D or L series.

❗ 13

a) **Stereoselectivity** means that in a chemical reaction one stereoisomer or a small group of stereoisomers of several possible stereoisomers is preferentially or even exclusively produced.

b) A reaction is termed **stereospecific** if the stereoisomeric starting materials are converted into stereoisomeric products. If the configuration of the starting materials is known, then the product of a stereospecific reaction can be predicted.

❶ 14

In the determination of the absolute configuration of cysteine it is important to note that the sulfur atom has a higher priority than the oxygen atom. The priority order of the atoms and groups directly attached to the chirality centre by application of the CIP rules is N > C(S,H,H)) > C(O,O,(O)) > H. Thus L-cysteine has the R configuration in contrast to other L-amino acids which have the S configuration.

❶ 15

a) The term **inversion** – depending on the context – has several meanings. It is most frequently used to describe the steric course of a substitution reaction when the arrangement of atoms or groups at a chirality centre is reversed relative to the substituted group.

Inversion is also used to describe the mutual interconversion of six-membered ring chair conformations.

Inversion is also used to describe the oscillation of the central atom of a trigonal pyramidal structure through the plane formed by the groups attached to this atom.

b) An achiral molecule is termed **prochiral** if it can be converted in a single transformation (chemical reaction) to a chiral molecule.

c) The **topicity** is the spatial relationship between constitutionally and configurationally identical (homomorphic) atoms or groups of atoms in a molecule. The groups or faces in a molecule are classified as either homotopic or heterotopic. The heterotopic groups or faces are further sub-divided into enantiotopic and diastereotopic groups or faces, and the topicity is denoted by the stereodescriptors *pro-R, pro-S, Re, Si, pro-E* and *pro-Z*.

❶ 16

There are two prochirality centres in butanone at carbon atoms 2 and 3.

! 17

The **relative configuration** of a molecule is the spatial arrangement of the atoms or groups in relation to other groups in the molecule. In contrast to the absolute configuration it is unchanged on reflection. A series of stereo-descriptors exist to describe the relative configuration, most of which can be applied to describe either pure enantiomers or racemates. For cyclic systems possessing two substituents the stereodescriptors *cis* and *trans* are used, while *r*, *c* and *t* are used if three or more substituents are attached to a ring. The orientation of substituents in bicyclic systems is denoted by the stereodescriptors *syn*, *anti*, *exo* and *endo*. For compounds with only two chirality centres the stereodescriptors *l* and *u* indicate the same and opposite absolute configuration of the chirality centres in the molecule, respectively. When the orientation of two groups, each attached to a chirality centre, in a molecule can be unequivocally represented in terms of a Fischer projection formula, the stereodescriptors *erythro* and *threo* can be used. For a pure enantiomer of unknown absolute configuration, the stereodescriptors *R**, *S** or *rel* are employed.

The term relative configuration is also used in the comparison of various types of compounds which differ only by one substituent at a stereogenic unit, e.g. the starting material and the product in a nucleophilic substitution reaction. In this instance relative configuration refers to either the group being exchanged or to the remaining groups.

! 18

The eclipsed conformation in which the two hydroxy groups are coincident at either end of the bond joining the two centres is referred to as synperiplanar (*sp*). If the group at the back of the C-C bond is rotated in a clockwise manner with respect to the observer, the staggered +synclinal (*+sc*) conformer is reached. Further rotation leads to the eclipsed anticlinal (*+ac*) conformation and then to the staggered antiperiplanar (*ap*) conformer. If the rotation is continued, then the –anticlinal and –synclinal conformations, respectively, are produced. Synclinal conformations are also called gauche conformations. All conformations in between eclipsed and staggered are known as skew or skewed conformations. A skewed conformation has the descriptor of the eclipsed or staggered conformation next to it.

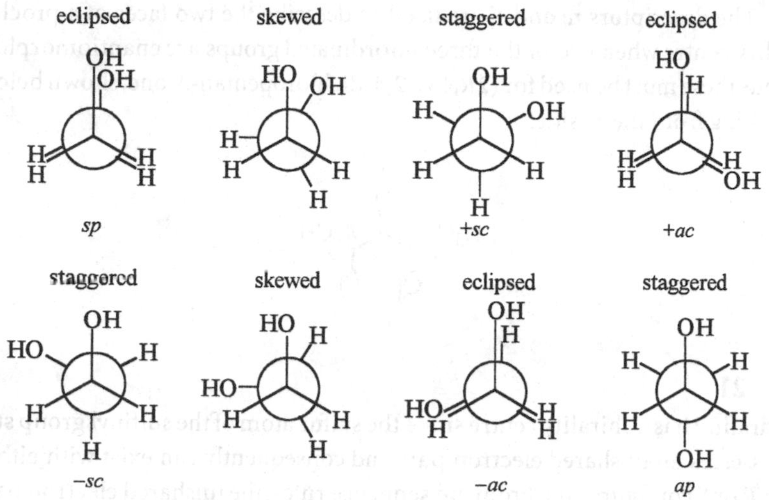

eclipsed	skewed	staggered	eclipsed
sp	*+sc*	*+ac*	

staggered	skewed	eclipsed	staggered
−sc	*−ac*	*ap*	

! 19

A **pseudochirality centre** exists on a tetrahedrally coordinated atom when two of the substituents are constitutionally identical but have the opposite chirality sense, i.e. they are enantiomorphic. An *R*-configured group has priority over an *S*-configured group. Thus it is possible to determine the configuration of a pseudochirality centre by application of the sequence rules. To distinguish a pseudochirality centre from a chirality centre the stereodescriptors *r* and *s* are used.

! 20

a) The descriptors *Re* and *Si* are used to describe the relationship of the two heterotopic faces of a trigonal planar prochirality centre with three constitutionally different groups. Accordingly, in (*R*)-3-chlorobutan-2-one these descriptors must be used. In the formula shown below the view is from the *Si* side.

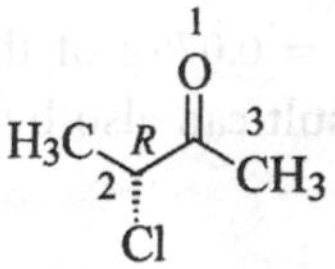

b) The descriptors *re* and *si* are used to describe the two faces of a prochirality centre when two of the three coordinated groups are enantiomorphic. Thus these must be used for (2*R*,4*S*)-2,4-dichloropentan-3-one, shown below looking from the *re* side.

21

Adrafinil has a chirality centre since the sulfur atom of the sulfinyl group still possesses an unshared electron pair and consequently can exist with either an *R* or *S* configuration. From the sequence rules, the unshared electron pair, which in the *S* enantiomer shown below is pointing towards the observer, has the lowest priority. The priority order of the groups attached to the sulfur atom is O > C(C,C,H) > C(C,H,H) > e⁻. The mirror image of this formula represents the *R* isomer. Note that sulfoxides with two different groups attached always have a chirality centre, since the sulfur atom cannot oscillate through the plane (i.e. there is no pyramidal inversion).

22

Since the original compound has an enantiomeric excess of 85 %, it contains 85 % of the pure enantiomer and 15 % of the racemate. Thus for a 1 g sample there is 150 mg of the racemate which must contain 75 mg of each enantiomer. Therefore 75 mg = 0.075 g of the (+)-enantiomer must have been separated. The same result can also be obtained using the following equation.

$$\% \ ee = \frac{|[E_1] - [E_2]|}{|[E_1] + [E_2]|} \cdot 100$$

Although this equation has two unknowns, it is solvable because of the limiting condition

$$[E_1] + [E_2] = 1$$

and this can be used as a second equation. In the solution shown below the units have been omitted for simplicity and it is assumed that E_1 represents the isomer in excess [i.e. the $(-)$-isomer in this instance] in order that the signs for the absolute values can be ignored.

$$85 = \frac{E_1 - E_2}{E_1 + E_2} \cdot 100$$

$$\frac{85}{100} = \frac{E_1 - E_2}{1}$$

$$0{,}85 = E_1 - E_2$$

$$E_2 + 0{,}85 = E_1$$

$$E_2 = E_1 - 0{,}85$$

$$E_2 = 1 - E_1 \quad \text{(limiting condition)}$$

$$2E_2 = 1 - 0{,}85$$

$$E_2 = \frac{1 - 0{,}85}{2} = 0{,}075$$

The last line of this method of solving the problem corresponds to the equation

$$[E_2] = \frac{1 - ee}{2}$$

which is a mathematical formulation of the rationale used to solve the problem given in words in the opening paragraph.

$(-)$-Ethyl lactate is the ethyl ester of $(+)$-lactic acid.

(+)-lactic acid   (−)-lactic acid ethyl ester

🛈 23

The isomers of butene are but-1-ene (with only a plane of symmetry, symmetry point group C_s), (E)-but-2-ene (with a horizontal plane of symmetry, a twofold axis of symmetry perpendicular to it, and a centre of symmetry, symmetry point group C_{2h}), (Z)-but-2-ene (with a twofold proper axis of symmetry and two planes of symmetry containing this axis, symmetry point group C_{2v}) and isobutene (2-methylpropene) with two mutually perpendicular planes of symmetry and on the line of intersection of these two planes there is a twofold axis of symmetry. The symmetry point group is therefore C_{2v}. These results can be verified using the flow chart in the appendix.

🛈 24

2-Methylcyclohexan-1-ol has two chirality centres and therefore $2^2 = 4$ stereoisomers are possible. A and B are the isomers with a *trans* configuration, and C and D are the *cis*-configured isomers. Isomers A and B are enantiomers as are C and D, while A and B are diastereomers of both C and D.

🛈 25

The formula is that of the R isomer of debropol. The priority order of the groups attached to the chirality centre is $Br > NO_2 > CH_2OH > CH_3$. Since the methyl group is already directed away from the observer, the configuration is readily deduced.

❗ 26

The groups of higher priority at each end of the double bond (the ester and isopropyl groups in this example) must lie on the same side, when the double bond in 2-cyano-3,4-dimethylpent-2-enoic acid methyl ester has the *Z* configuration.

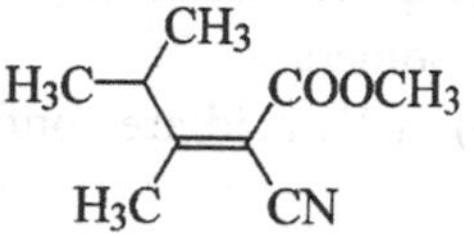

$$H_3C-\overset{\displaystyle CH_3}{\underset{\displaystyle H_3C}{|}}C=\overset{\displaystyle COOCH_3}{\underset{\displaystyle CN}{}}$$

❗ 27

Fudosteine is a derivative of L-cysteine (see question 14) and therefore has the *R* configuration. The order of priority of the atoms and groups attached to the chirality centre by application of the CIP rules is N > C(S,H,H) > C(O,O,(O)) > H.

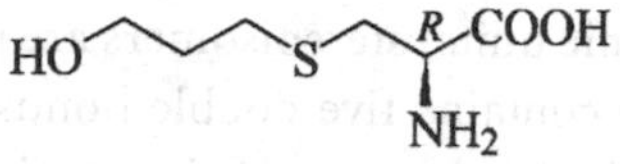

$$HO\diagup\diagup S\diagup\overset{R}{\underset{NH_2}{|}}COOH$$

❗ 28

1-Bromopenta-1,2,3-triene is a cumulene with an odd number of double bonds and consequently *E* and *Z* isomers are possible, if at each end of the unsaturated system there are two different groups (one of which can be hydrogen). Since the *Z* configuration is specified in the question, the bromine atom and the methyl group must lie on the same side of the planar unsaturated system. Therefore (*Z*)-1-bromopenta-1,2,3-triene has the formula shown below.

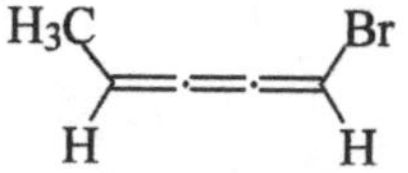

$$\overset{\displaystyle H_3C}{\underset{\displaystyle H}{}}C==C\overset{\displaystyle Br}{\underset{\displaystyle H}{}}$$

❗ 29

a) (*E*)-1-Bromopropene and (*Z*)-1-bromopropene are configurational isomers (diastereomers).

b) L-Alanine [(*S*)-2-aminopropanoic acid] and β-alanine (3-aminopropanoic acid) are constitutional isomers.

c) Lactic acid (2-hydroxypropanoic acid) and 3-hydroxybutanoic acid differ in constitution but are not isomers.

d) (−)-Lactic acid and (+)-lactic acid are configurational isomers (enantiomers).

e) 1-Chloropropene and 2-chloropropene are constitutional isomers.

f) *cis*-2-Chlorocyclohexanol and *trans*-2-chlorocyclohexanol are configurational isomers (diastereomers).

❗ 30

According to the formula for the calculation of the theoretical number of stereoisomers (configurational isomers), $x = 2^n$, $2^4 = 16$ (where $n = 4 =$ the number of stereogenic units) stereoisomers are possible for alitretinoin. Although the molecule contains five double bonds only those in the side chain can be considered, since because of ring strain the double bond in the six-membered ring can only have the Z configuration.

❶ 31

a) Ethanol: no configurational isomers

b) Butan-2-ol: two enantiomers

c) Glycerol: no configurational isomers

d) 2,3-Dibromobutane: three configurational isomers

e) Acetone oxime: no configurational isomers

f) Pent-3-en-2-ol: four configurational isomers

g) Pentane-2,3-diol: four configurational isomers

h) Pentane-2,4-diol: three configurational isomers

i) 3-Bromobutan-2-ol: four configurational isomers

j) But-2-enoic acid: two diastereomers

k) 4-Ethylhepta-2,5-diene: four configurational isomers; Both double bonds here can have the same configuration (E,E or Z,Z) or may have a different configuration. In the latter case carbon atom 4 will be a chirality centre and therefore two enantiomers exist. For the determination of the appropriate stereodescriptor by the CIP rules, the Z-configured group is ranked higher than the E-configured group.

l) Hexa-2,3,4-triene: two diastereomers (E and Z isomers)

32

In the chair form shown, the numbering of the ring atoms is clockwise. To transfer the formula into a Fischer projection formula is a relatively simple task. All groups in the ring which are below the plane of the ring are located on the right hand side in the Fischer projection formula. Thus the hydroxy group at the anomeric centre is also on the right hand side. Since it is on the same side of the oxygen atom as the anomeric reference atom, which in galactose as in most other carbohydrates is the same as that which determines whether it belongs to the D or L series, the compound in the current example has the α configuration. The formula therefore represents α-D-galactopyranose.

33

a) Examples are compounds with a chirality centre and no other stereogenic units, e.g. lactic acid.

b) Examples here include almost all compounds with more than one chirality centre, e.g. tartaric acid, cholesterol and octahedral coordination compounds with more than three different ligands.

c) Examples include 1,4-dichlorocyclohexane and all compounds with stereogenic units not generating chirality such as double bonds, e.g. maleic acid and square planar coordination compounds without chiral ligands.

❗ 34

There are six isomers of difluorocyclobutane (see below). The vicinal di-substituted isomers **B** and **C** (both with a twofold proper axis of symmetry, symmetry point group C_2) are chiral and are enantiomers of each other. The *cis*-configured compound **D** (with a plane of symmetry, symmetry point group C_s) is achiral and is a meso compound. The compounds **A** and **F** (both with two planes of symmetry and on the line of intersection of both planes a twofold axis of symmetry, symmetry point group C_{2v}) and **E** (with a plane of symmetry, a twofold axis of symmetry perpendicular to it and a centre of symmetry, symmetry point group C_{2h}) are all achiral. These results can be verified from the flow chart given in the appendix.

❗ 35

The epimers of (2R,3S)-bicyclo[2.2.1]heptane-2,3-diol are (2R,3R)-bicyclo-[2.2.1]heptane-2,3-diol and (2S,3S)-bicyclo[2.2.1]heptane-2,3-diol. Note that epimers differ in the absolute configuration at only one chirality centre. As a consequence of structural constraints, only two epimers are possible since the bridge can only be formed by a *cis* linkage.

❗ 36

a)

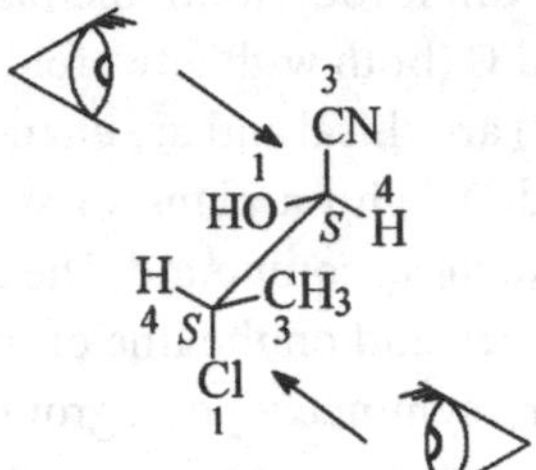

b)

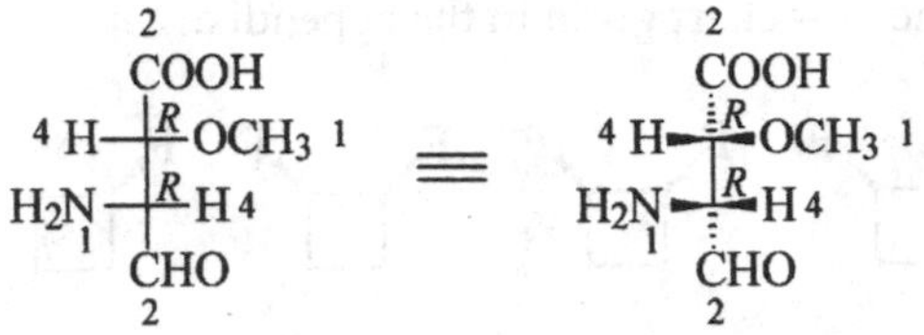

c)

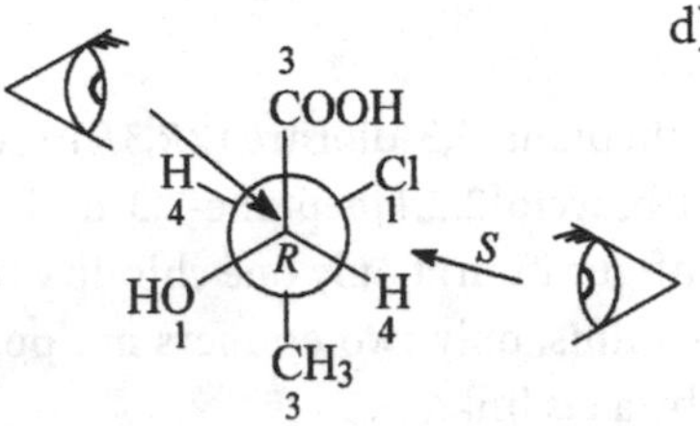

d)

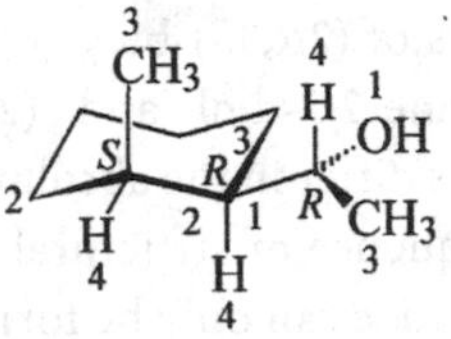

❗ 37

a) The simplest approach is to first of all determine the absolute configuration from both formulae. The chirality centre in the compound drawn as a Fischer projection formula has the S configuration, whilst the absolute configuration of the second compound is R. The two compounds are therefore enantiomers of each other and are L- and D-alanine, respectively. Alternatively, it would also be possible to transform the Fischer projection formula of the first compound into a formula with wedged bonds and then rotate one of the formulae until it is obvious that they are mirror images.

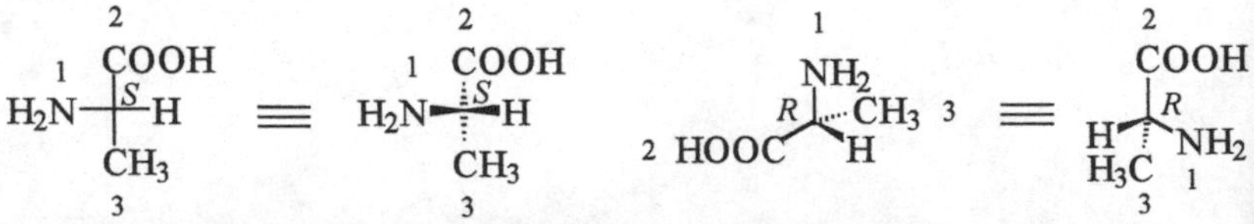

b) In this example the best approach is to rotate one of the formulae given through 120° and check whether it is superimposable with the other formula. This is not the case. The next check is to see whether one of the formulae can be converted into the other by reflection. This is possible and therefore the two compounds represented by the formulae are enantiomers of each other. As a final check, the configuration of the chirality centres should be determined since these must be inverted on reflection.

c) Since it is relatively time consuming to rotate each single bond in the Fischer projection formula to arrive at a zigzag projection, the simplest approach is – after a check of the constitution – to determine the configuration of all the chirality centres and then compare the results. The two structures are diastereomers of one another, and are the epimers D-xylose and L-arabinose whose structures differ at carbon atom 4 only.

D-xylose   L-arabinose

d) The two coordination compounds differ in the positions of the bromine atoms and the carbonyl ligands, and therefore are diastereomers.

! 38

a) (*E*)-1,2-Dichloroethene and (*Z*)-1,2-dichloroethene are diastereomers.

b) (+)-Tartaric acid and *meso*-tartaric acid are diastereomers.

c) (1*R*,2*S*)-Cyclohexane-1,2-diamine and (1*R*,2*R*)-cyclohexane-1,2-diamine are diastereomers.

d) (1*S*,2*S*)-Cyclohexane-1,2-diamine and (1*R*,2*R*)-cyclohexane-1,2-diamine are enantiomers.

e) α-D-Glucopyranose and β-D-glucopyranose are diastereomers. They are also epimers and anomers.

α-D-glucopyranose   β-D-glucopyranose

f) α-D-Mannopyranose and α-L-mannopyranose are enantiomers.

α-D-mannopyranose   α-L-mannopyranose

! 39

The reaction with Mosher's acid chloride is a normal esterification reaction with retention of configuration of the chirality centres of both reactants. Since, however, the chlorine atom in the acid chloride is replaced, nevertheless the descriptor for the Mosher's acid part of the product will change.

❗ 40

a) Acetylene has a linear structure. It possesses an infinite axis of symmetry along the molecular axis and an infinite number of perpendicular twofold axes of symmetry intersecting in the centre of the molecule where there is also a centre of symmetry. In addition, there is a plane of symmetry perpendicular to the main rotational axis and there are innumerable (vertical) planes of symmetry containing the molecular axis. These symmetry elements lead to the point group $D_{\infty h}$.

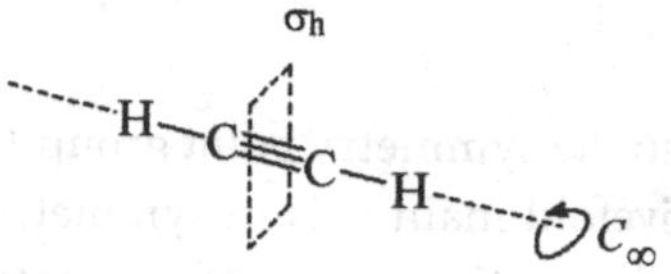

b) Hydrogen peroxide is not a flat molecule but angular with a dihedral angle slightly greater than $90°$. It possesses a single symmetry element, i.e., a C_2 axis, and therefore belongs to the point group C_2. This is easier to visualise when the molecule is viewed in the Newman projection.

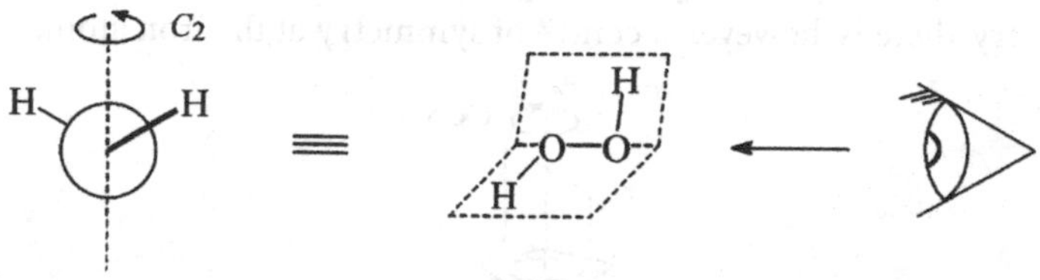

c) White phosphorus consists of discrete P_4 molecules. Sets of three phosphorus atoms in the P_4 molecule form equilateral triangles. A structural formula is most easily drawn if the phosphorus atoms occupy the corner positions of opposing diagonals in the faces of a cube. This tetrahedral arrangement leads to the symmetry point group T_d. The symmetry elements comprise of four C_3 axes passing through each phosphorus atom and through the centre of the opposite face (for simplicity only one is shown in the diagram) and six planes of symmetry which lie in the planes containing the threefold axes of symmetry. The planes of symmetry each contain two corners of the tetrahedron and bisect the opposite edge (P-P bond). There are three twofold axes of symmetry which occupy the same position as three S_4 axes and pass through the centres of opposing edges of the tetrahedron.

The two C_2 and S_4 axes not shown (for clarity) are perpendicular to the one indicated in the diagram.

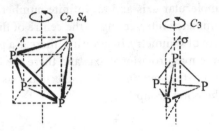

d) Ferrocene belongs to the symmetry point group D_{5d}. The symmetry elements present are a fivefold main axis of symmetry C_5, perpendicular to which are five C_2 axes, five vertical planes of symmetry σ_d which each bisect the angle between two neighbouring C_2 axes, and an S_{10} alternating axis of symmetry. The S_{10} axis is parallel to the principal C_5 axis and consists of a C_{10} symmetry operation (rotation through $36°$) followed by reflection in a perpendicular plane. Since both cyclopentadienide rings are skewed (the carbon atoms in the two rings are staggered), there is no horizontal plane of symmetry, there is, however, a centre of symmetry at the iron atom.

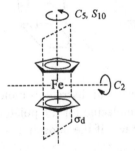

e) Twistane belongs to the symmetry point group D_2. It has apart from three mutually perpendicular twofold axes of symmetry, no other symmetry elements. It is therefore chiral and exists in two enantiomeric forms.

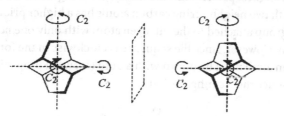

● 41

The chirality centre at position 3 in the bicyclic system of vedaclidine has an S configuration. The order of priority of the atoms and groups by application of the CIP rules is C(N,(N),C) > C(N,H,H) > C(C,C,H) > H. As the hydrogen atom is directed downwards, the observer must determine the sequence of the remaining groups by viewing the molecule from above. In the projection shown it is possible to determine the sequence directly. Note that the bridgehead atoms 1 and 4 of the quinuclidine ring are not chirality centres since both are attached to two identical groups.

❶ 42

Linezolid contains a chirality centre in the oxazolidine ring. The ring oxygen atom has the highest priority and the methylene group attached to the nitrogen atom with two neighbouring carbon atoms has a higher priority than the methylene group attached to the nitrogen atom with only one neighbouring carbon atom. However, since the sequence is clockwise in the formula given, the hydrogen atom must be drawn with a bold (forward projecting) wedged bond to produce the *S* configuration.

❶ 43

a) (*Z*)-1,2-dibromo-1-chloro-2-iodoethene

b) (*R*ₐ)-1,3-dichloroallene [(*R*)-1,3-Dichloroallene is also a possibility. However, it is not recommended to omit the index a, which indicates that the stereodescriptor describes the configuration of a chirality axis.]

c) (4*S*)-4-chloro-2-methyloctane

d) (3*S*,4*R*)-4-hydroxy-3-methylpentanoic acid

❗ 44

a) In order to establish the relationship between the isomers, the absolute configuration must be determined. In molecule **B** this is immediately obvious since the atom of lowest priority (the hydrogen atom) is directed away from the observer. In order to determine the configuration of molecule **A**, the molecule must first of all be rotated in a clockwise direction about a vertical axis until the hydrogen atom is directed away from the observer. By application of the sequence rules, both compounds have the R configuration and are therefore identical.

A ≡ **B**

b) Structure **A**, with both methyl groups in axial positions, must first of all be ring flipped to the conformer in which these groups occupy equatorial positions. This operation does not change the configuration of the chirality centres. The resulting formula can be transformed to formula **B** by rotating it through 180°. Finally determination of the configuration of the chirality centres proves that all three formulae represent the same molecule.

c) Both Haworth projections represent the same molecule. Clockwise rotation of the first formula through 120° leads to the second formula.

d) The first formula must be converted into the corresponding chair form. Care is necessary to bring the methyl groups into the correct positions in order to end up with the same configuration. If necessary, the ring is then flipped. Comparison with the chair form which was given in the question shows that the molecules are enantiomers. (This method is the shortest way to solve the problem but not the only one. Another approach is, for example, to compare directly the absolute configuration of all the chirality centres determined from both formulae.)

❗ 45

a)

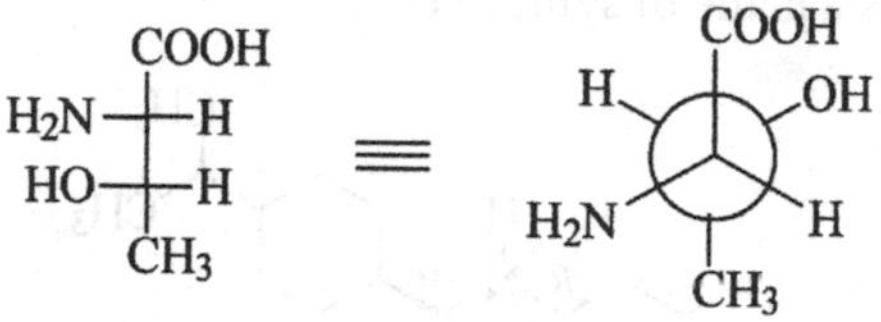

b)

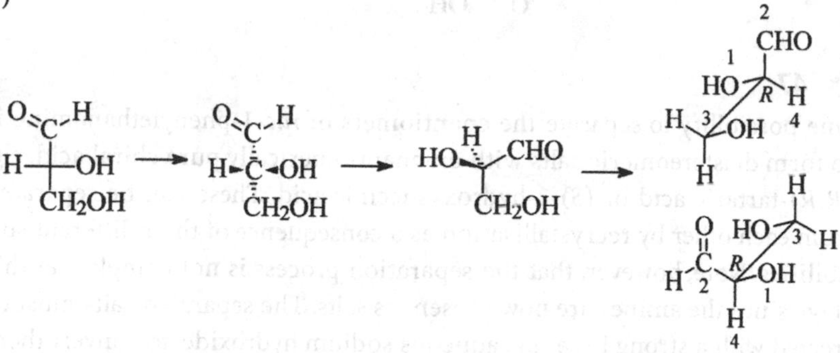

c)

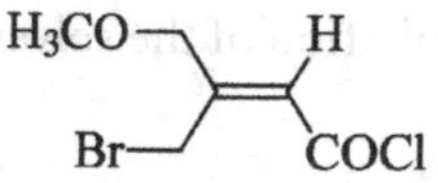

d)

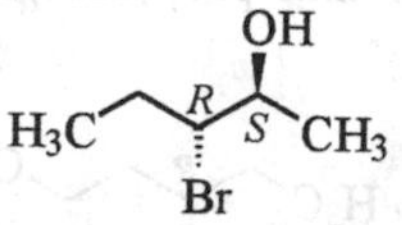

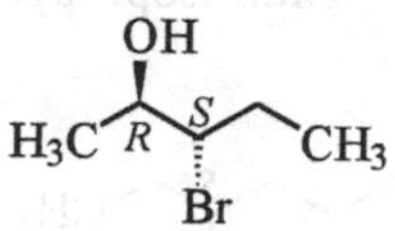

e)

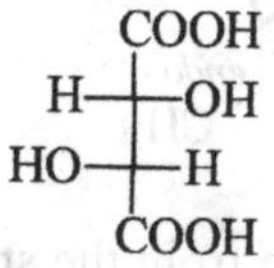

f)

Since it is impossible to derive the absolute configuration from the steric descriptors enantio, exo, syn, and endo, the name can be used for the pure enantiomers as well as the racemate.

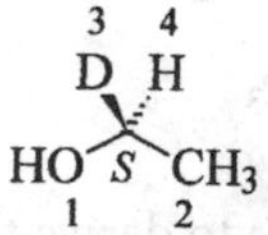

❗ 46

Nateglinide has an *R*-configured chirality centre and is derived from D-phenyl-alanine. The substituted *trans*-configured cyclohexyl residue is achiral since it has a plane of symmetry.

❗ 47

One possibility to separate the enantiomers of *rac*-1-phenylethanamine is to form diastereomeric salts with an enantiomerically pure chiral acid, e.g. (*R,R*)-tartaric acid or (*S*)-2-hydroxysuccinic acid. These can be separated from each other by recrystallisation as a consequence of their different solubilities. Note, however, that the separation process is not complete at this stage since the amines are now present as salts. The separated salts must be treated with a strong base, e.g. aqueous sodium hydroxide, to convert them back to the free amines which can then be extracted into an organic solvent. After drying the extract distillation of the solvent leaves the pure amine.

❗ 48

7-*syn*-Ethyl-5-*endo*-isopropyl-6-*exo*-methyl-7-*anti*-propylbicyclo[2.2.1]-hept-2-ene

Since it is impossible to derive the absolute configuration from the stereo-descriptors *endo, exo, syn* and *anti*, the name can be used for the pure enantiomers as well as the racemate.

❗ 49

Diamminedichloridoplatinum(II) can exist as the two isomers shown below. Isomer **A** is the cytostatic agent cisplatin and usually described by the stereo-descriptor *cis*, whilst its cytostatically inactive diastereomer **B** is assigned the stereodescriptor *trans*.

More generally applicable are the systematic descriptors for inorganic coordination compounds based on the CIP system. These consist of a polyhedral symbol, *SP*-4 in the current example (*SP* for square planar and 4 for the coordination number), and a symbol for the configuration index. In square planar compounds the latter is a one digit number. This gives the priority number for the coordinating atom (ligand) which is *trans* to the highest ranking coordinating atom (ligand). Thus isomer **A** has the descriptor *SP*-4-2. It has two planes of symmetry and a twofold axis of symmetry formed by the intersection of the two planes of symmetry and belongs therefore to the point group C_{2v}. Isomer **B** is (*SP*-4-1)-diamminedichloridoplatinum(II) and belongs to the point group D_{2h}. There are three mutually perpendicular twofold axes of symmetry, three mutually perpendicular planes of symmetry, and a centre of symmetry which is equivalent to a twofold alternating axis of symmetry.

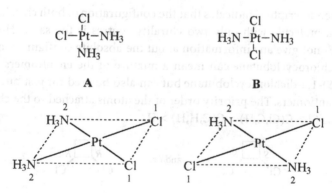

❗ 50

1,4-Dimethylbicyclo[2.2.1]heptan-2-ol contains three chirality centres. Therefore the total number of isomers which is theoretically possible (x) can be calculated using the formula $x = 2^n$: $2^3 = 8$. However, since the bridge can only be formed by *cis* fusion due to geometrical constraints, both the bridgehead methyl groups must be *cis* to each other. This halves the number of stereoisomers to two pairs of diastereomeric enantiomers, i.e. 1 and *ent*-1 and 2 and *ent*-2. Compound 1 is a diastereomer of both 2 and *ent*-2; *ent*-1 is diastereomeric with both 2 and *ent*-2; 2 is a diastereomer of both 1 and *ent*-1; and *ent*-2 is diastereomeric with both 1 and *ent*-1. Compounds 1 and 2 are epimers as are *ent*-1 and *ent*-2.

❗ 51

The stereodescriptor *l* indicates that the configuration of both chirality centres in a molecule with only two chirality centres is the same. However, this does not give any information about the absolute configuration. Thus *l*-1,2-dichlorocyclobutane can mean a mixture of the enantiomers (*R,R*)- and (*S,S*)-1,2-dichlorocyclobutane but can also be used for just one of the pure enantiomers. The priority order of the atoms attached to the chirality centres is Cl > C(Cl,C,H) > C(C,H,H) > H.

❗ 52

The first step here is to draw formulae of both isomers of cromakalim and then to determine the absolute configuration of both compounds using the CIP rules. The two compounds are the $3S,4R$ isomer and the $3R,4S$ isomer, and are enantiomers. The INN for the pure $3S,4R$ enantiomer is levcromakalim.

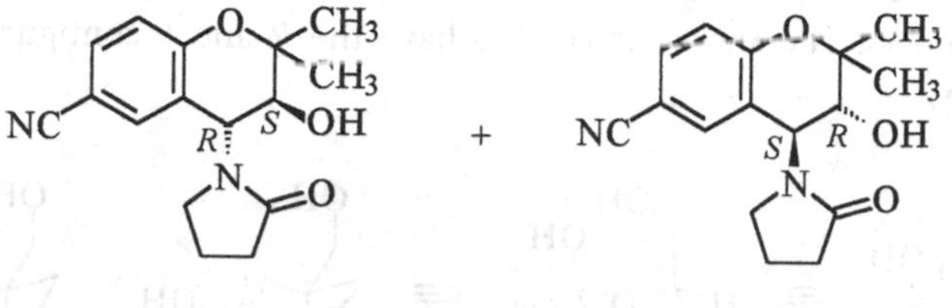

levcromakalim

❗ 53

a) Both formulae represent the same compound which has the S configuration. If the compound represented by the formula on the left is viewed from the right hand side along the molecular axis, then the projection shown on the right is produced. Please note, however, that the Newman projection formula on the right, in the absence of additional information, could similarly represent a different structure because the number and type of atoms located on the axis are not indicated.

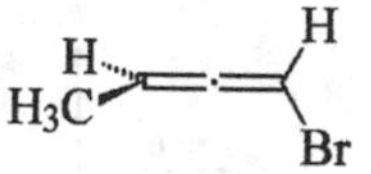

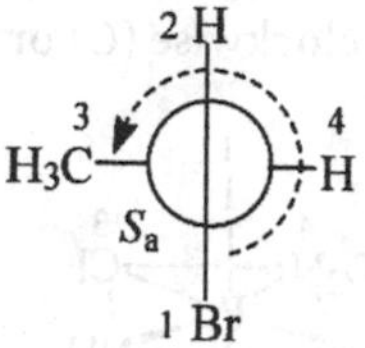

b) The first step here is to convert the Haworth projection into a chair form. Two chair conformations are possible, i.e., $^1C_4$ and $^4C_1$ and these must both have the identical absolute configuration. In order to avoid any errors during the transformation of formulae it is best to determine the configuration at both chirality centres.

Comparison of the $^4C_1$ configuration with the chair form given in the question leads to the conclusion that the compounds are anomers and hence diastereomers. The anomeric centres have the R and S configuration, respectively.

$^1C_4$ $^4C_1$

c) Both complexes differ in the positions of the nitrito and ammine ligands. The remaining ligands lie in the same plane, thus the compounds are enantiomers. The same result can also be obtained by determining the stereodescriptors for the complexes. This is accomplished by examining the four ligands in the plane perpendicular to the axis containing the ligand of highest priority, and viewing from the ligand of highest priority, to see whether their arrangement according to the priority order derived from the CIP system is clockwise (C) or anticlockwise (A).

OC-6-25-A   OC-6-25-C

d) Both coordination compounds contain three identical ligands in the same plane. If either of the formulae is rotated each time through 90° around two mutually perpendicular axes, the other formula is produced, i.e., both formulae represent the same compound. The same solution can be obtained by determining the stereodescriptors for the complexes. The stereodescriptor comprises the polyhedral symbol OC-6 and the configuration index. For both complexes this is the digit 2 twice, i.e., the stereodescriptor is OC-6-22 since all three groups of higher priority lie opposite a group with the priority number of 2. (The stereodescriptor fac was previously used to describe such complexes since both groups of three similar ligands each occupy the corners of a face of the octahedron.)

54

The compound shown has an R-configured chirality centre [the priority order of the groups attached to the chirality centre is $C(P,H,H) > C(C,C,(C))$ $> C(C,H,H) > H$] and a Z-configured double bond. There are therefore three stereoisomers which have the opposite configuration either at one or both stereogenic units. Thus, including the isomer shown there are a total of four stereoisomers with this constitution, i.e., the R,Z-, R,E-, S,Z- and the S,E-configured molecules. (Compounds of this type have been synthesised as potential antiviral nucleotide analogues and it has emerged that only the Z isomers show any antiviral activity [1].)

55

❗ 56

(R)-1-Bromobuta-1,2-diene is a cumulene with an even number of double bonds and, if (as is the case in this example) there are two different groups attached at both ends of the cumulene system, it will possess a chirality axis. The absolute configuration can be determined, if the molecule is viewed like in a Newman projection along the chirality axis which coincides with the axis of the double bonds. The viewing direction is chosen at random. The substituents at one end will lie in a vertical direction and at the other end in a horizontal direction. Next the substituents closer to the viewer are connected with a clockwise curve, since an R configuration is required, starting from the higher ranked substituent (the bromine atom or the methyl group depending on which end the molecule is viewed from) and the curve continued by another 90° where the other substituent is placed.

To denote a stereodescriptor for a chirality axis a subscript a is used (not italicised). Thus in this example the compound is (R_a)-1-bromobuta-1,2-diene. Although it has been common practice to omit the a in the stereodescriptor, for clarity this is not recommended and should not be done.

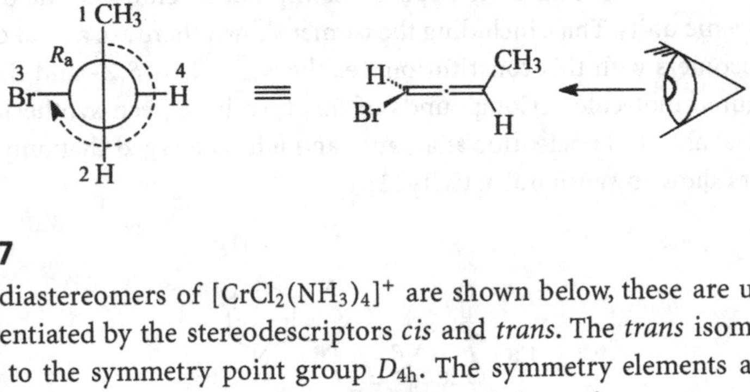

❗ 57

Both diastereomers of $[CrCl_2(NH_3)_4]^+$ are shown below, these are usually differentiated by the stereodescriptors *cis* and *trans*. The *trans* isomer belongs to the symmetry point group D_{4h}. The symmetry elements are the main fourfold axis of symmetry C_4, a horizontal plane of symmetry σ_h (perpendicular to the C_4 axis), four C_2 axes also perpendicular to the C_4 axis and four planes of symmetry σ_v the intersection of which is the main axis of symmetry. The *cis* isomer belongs to the symmetry point group C_{2v}. The associated symmetry elements are a C_2 axis and two vertical planes of symmetry σ_v intersecting at the C_2 axis. Verify this using the flow chart in the appendix.

Since the descriptors *cis* and *trans* are not generally applicable for octahedral coordination compounds, systematic descriptors based on the CIP system should be used in their place. These consist of the polyhedral symbol, in this example OC-6 (OC for octahedral and 6 for the coordination number), together with the configuration index. For octahedral compounds the latter consists of two digits. The first indicates the priority number of the coordinated atom (ligand) *trans* to the highest ranking coordinated atom (ligand). In the *cis* isomer this is 2, and 1 in the *trans* isomer. The second digit is determined in the same way for the plane perpendicular to the reference axis (main axis) of the octahedron. The *cis* isomer has thus the descriptor OC-6-22. The *trans* isomer is (OC-6-12)-tetraamminedichloridochromium(III).

58

There are $2^3 = 8$ possible isomers with the cyclothiazide constitution. The compound has in fact four chirality centres and therefore 16 isomers might have been imagined. However, since the configuration at the bridgehead atoms 1 and 4, for reasons of geometrical constraints (the bridge can only be formed by *cis*-fusion), cannot be independently varied, there are only eight isomers. Cyclothiazide, although now almost obsolete, was used as a mixture of isomers.

● 59

If (R_a)-1,3-dichloroallene is viewed along the axis of the double bonds, then the groups 1, 2 and 3 in the priority order according to the CIP rules are arranged in a clockwise sense to the observer. Alternatively, only the chlorine atoms need to be considered (the groups of higher priority at either end of the compound). The curve from the chlorine atom closer to the observer to the other chlorine atom has a counter-clockwise sense. The resulting helicity is described by the stereodescriptor M. Therefore (R_a)-1,3-dichloroallene is identical with (M)-1,3-dichloroallene. In either approach it is irrelevant to the observer from which end the molecule is viewed.

● 60

The stereodescriptor u denotes that two chirality centres are present the configuration of which is unlike. One of these chirality centres is due to the unshared electron pair on the substituted pyramidal sulfur atom. The unshared electron pair is given the lowest priority in determining the configuration.

❶ 61

a) The two faces of the double bond of butanone are enantiotopic. The view onto the formula shown is from the *Re* side.

$$\underset{H_3C}{\overset{3}{}}\overset{\overset{1}{O}}{\underset{}{\|}}\overset{2}{}CH_3$$

b) Since the compound already possesses a chirality centre (at position 2 of the 2-chloropropanoic acid residue), both faces of the carbonyl group are diastereotopic. The view onto the formula shown is from the *Re* side.

$$H_3C\overset{1}{\smile}O\overset{\overset{2}{O}}{\underset{}{\|}}\overset{3}{\underset{\underset{Cl}{\overset{*}{|}}}{}}CH_3$$

c) Since both faces of acetone are identical, they are homotopic.

d) The two ends of the double bond of bromoethene must be considered separately. At the bromine-substituted end, the two faces are enantiotopic. The view onto the formula shown is from the *Si* side. At the other end of the double bond the two groups are identical (hydrogen atoms), the two faces at this carbon atom are therefore homotopic.

$$\underset{H}{\overset{H}{}}\underset{2}{}\overset{1}{\underset{3}{}}\overset{Br}{\underset{H}{}}$$

❗ 62

a) Maleic acid reacts first with molecular bromine with loss of a bromide ion and the formation of a cyclic bromonium ion. In a second step the nucleophilic bromide ion can attack the cyclic bromonium ion. This attack proceeds as in a normal S_N2 reaction stereospecifically with inversion. Since there is equal probability of attack by the bromide ion at C2 and C3 of the *meso*-bromonium ion, the product is the racemate of (2R,3R)- and (2S,3S)-dibromosuccinic acid. It should be noted that the above reaction only proceeds in this manner in the absence of light and in the cold.

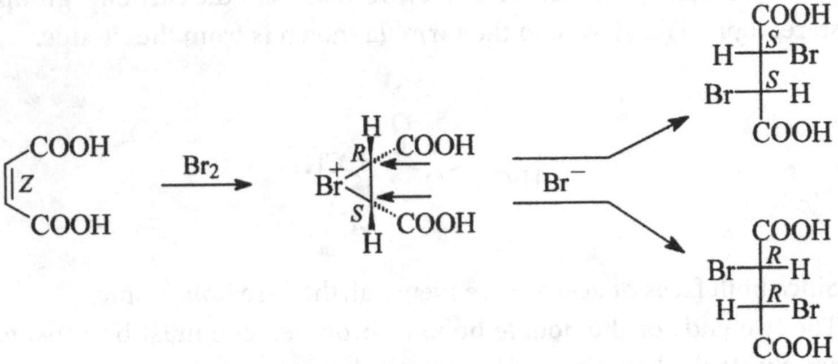

b) Under analogous conditions, fumaric acid yields two enantiomeric cyclic bromonium ions which have the R,R and S,S configuration, respectively. The subsequent ring opening of the three-membered ring by the nucleophilic attack of the bromide ion is independent on which of the bromonium ions is attacked. In either case the same product, (2R,3S)-dibromosuccinic acid (*meso*-2,3-dibromosuccinic acid), is produced, since nucleophilic attack always leads to inversion at one of the chirality centres. The two separate formulae shown below for the product are interconvertible by rotation through 180°.

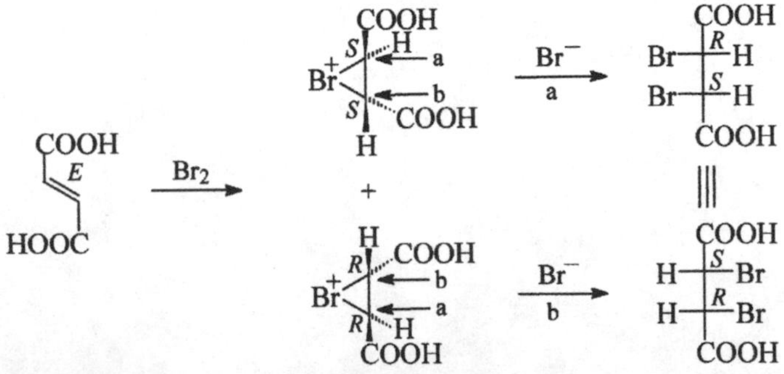

❗ 63

Lumefantrine contains two stereogenic units, i.e., a Z-configured double bond and a chirality centre whose configuration in the formula given is unspecified. As a result the formula represents two compounds (enantiomers) with R,Z and S,Z configuration, respectively. Both are present in the racemic pharmaceutical. There are formally $2^2 = 4$ isomers possible with this constitution; in addition to the isomers of lumefantrine there are also two others with an E-configured double bond.

❗ 64

Although there are two enantiomers of *trans*-1,2-dibromocyclopentane, there is only one epimer of it. Inversion of one chirality centre in either of the two enantiomers always leads to the same *cis*-configured compound.

ⓘ 65

It is initially best to draw the structure of (R)-2-methylbutane-1-thiol as a zigzag projection. The hydrogen atoms can then be added (this can be done either imaginary or actually). The reference group at position 1 is the sulfanyl group. At position 2 there are three different groups and the highest priority group here is the reference group. By looking along the C1-C2 bond, the Newman projection shown is obtained.

ⓘ 66

meso-Tartaric acid is initially best drawn as a Fischer projection formula which represents the synperiplanar conformation. This is then easily converted into a sawhorse projection. The desired anticlinal conformation is then obtained by rotating one half of the molecule through 120° about the central carbon-carbon bond. Note that there are two enantiomeric anticlinal conformations.

❗ 67

There are four stereoisomers of 4-*sec*-butylcyclohexanol. A chirality centre is present in the molecule and, in addition, *cis/trans* isomerism is possible in the ring. The four isomers are hence (*R*)-*trans*-, (*S*)-*trans*-, (*R*)-*cis*- and (*S*)-*cis*-4-*sec*-butylcyclohexanol.

❗ 68

Although there are formally six possible stereoisomers of an octahedral complex of general formula $Ma_2b_2c_2$, for dichlorido(diazane)bis(triphenylphosphane)cobalt(1+) only four isomers occur, since an arrangement of the two nitrogen atoms of the hydrazine *trans* to each other in the complex is not possible. The two OC-6-32 isomers are chiral, they are enantiomers and have the *A* and *C* configuration, respectively. Both the other two isomers are achiral diastereomers.

OC-6-13   OC-6-33   OC-6-32-*A*   OC-6-32-*C*

❗ 69

In order to determine the conformation, the substitution pattern at carbon atoms C2 and C3 must first of all be established. Three different groups are attached to each. Hence, the groups of highest priority on each are taken as the reference groups (shown bold in the diagram below), thus this must be the antiperiplanar conformation. In order to transform the formula into a Fischer projection formula, the part of the formula which is closer to the observer is rotated until an eclipsed conformation is obtained in which the groups forming the main chain are oriented synperiplanar to each other.

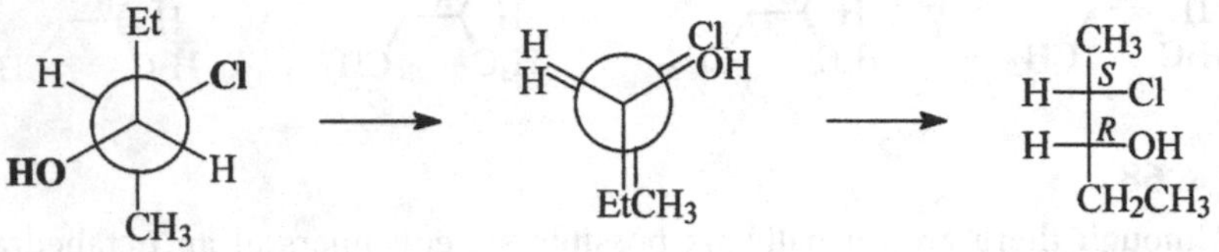

❗ 70

6,6'-Dibromobiphenyl-2,2'-dicarboxylic acid has a chirality axis, since there is restricted rotation due to the substitution pattern in the phenyl rings. Thus, depending on the orientation of the two rings, the plane of linear polarised light will either be rotated to the left or the right. In order to determine the configuration of the chirality axis, the molecule is viewed in a projection along this axis from either of the two sides. The atoms or groups of atoms in the ring closer to the observer are first considered, then those in the ring behind. In this case the compound has the R_a configuration.

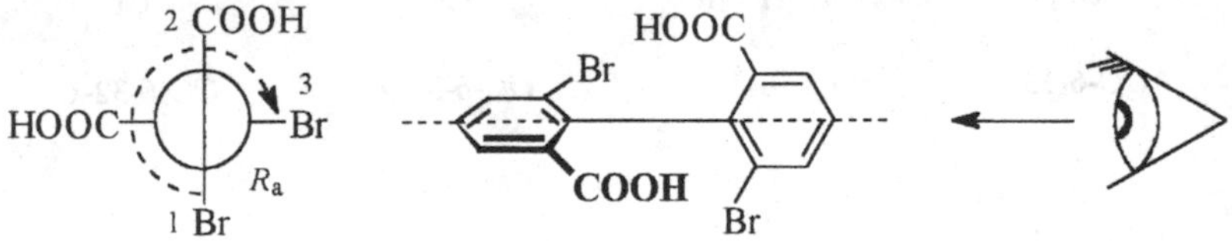

❗ 71

Amlodipine is chiral. Both halves of the heterocycle contain different substitutents, and consequently C4 is a chirality centre. Therefore two enantiomers exist, which differ in configuration at this centre in the 1,4-dihydropyridine ring. In determining the priority order according to the CIP system it soon becomes evident that in the first sphere the three groups to be considered are all carbon atoms attached to which are again three carbon atoms each, and only in the third sphere is it possible to determine the substituent of highest priority, i.e., the 2-chlorophenyl group. The priority order of the two remaining groups can only be determined in the fourth sphere, where the oxygen atom of the substituent in position 2 has precedence over the hydrogen atom of the methyl group in position 6. Note that the difference in the two ester groups plays no role in determining the priority order, because although they appear in the higher ranking branches of the digraph, the difference would only be decisive in the fifth sphere. The configuration would also be unaltered, if the positions of the two ester groups were interchanged, although in this instance constitutional isomers would result.

❗ 72

In 2*t*,3*c*-dichlorocyclohexan-1*r*-ol only the relative configuration of the substituents is known. Two enantiomers (shown below) exist with the absolute configuration 1*R*,2*S*,3*S* and 1*S*,2*R*,3*R*.

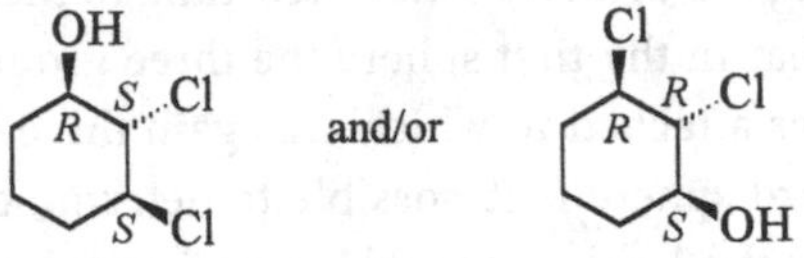

❗ 73

Only a plane of symmetry σ is present in 2-methylhex-3-yne, thus it belongs to the point group C_s. The methyl group on carbon 5 has free rotation and can lie in the plane of symmetry.

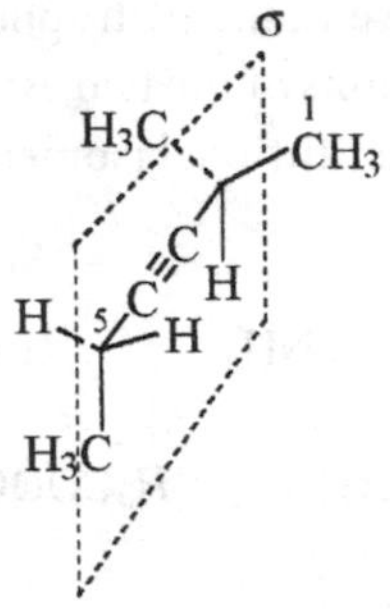

❗ 74

(S_a)-6-Aminospiro[3.3]heptan-1-ol possesses a chirality axis. The substituents at positions 2 and 6 lie in planes perpendicular to each other. The absolute configuration can be determined by viewing the molecule along the axis containing atoms 2, 4 and 6. The molecule can be viewed along this axis from either side. The substituents at one end will lie in a vertical direction and at the other end in a horizontal direction. Next the substituents closer to the viewer are connected with a counter-clockwise curve starting from the higher ranked substituent (the amino or the hydroxy group depending on which end the molecule is viewed from) and the curve continued by another 90° where the other substituent is placed.

Since *trans*-2-bromo-4-chlorocyclobutanone is chiral, the reaction of both enantiomers must be considered. For both enantiomers hydride attack from the *Re* side leads to a product with an *S* configuration at C1, whilst if attack takes place from the *Si* side a product with an *R* configuration at C1 is obtained.

❗ **76**

There are two stereoisomers of bis(2-aminoethanethiolato-*N*,*S*)nickel(II). These have the stereodescriptors (*SP*-4-1) and (*SP*-4-2). (The descriptors *trans* and *cis* are equally unambiguous here. Their use is, however, not recommended, since there are other square planar coordination compounds where these descriptors cannot be used.)

SP-4-1   *SP*-4-2

❗ 77

(2R,3R,4R)-3-Chloro-4-isopropyl-2-methylcyclohexanone can adopt two chair conformations of completely different energy contents. In one conformer all the substituents occupy equatorial positions, whilst in the other they are all situated in axial positions. Due to unfavourable 1,3-diaxial interactions, especially between the isopropyl and methyl group, the energy content of the latter conformer is raised significantly. In the conformer with the substituents in equatorial positions, only the gauche interactions of the substituents need be considered. These are however, much smaller by comparison and are of the same order as the gauche interactions of axial substituents with ring atoms. The conformer with the substituent groups equatorial will therefore have the lower energy content.

❗ 78

(Z)-Pent-2-ene is less stable than (E)-pent-2-ene and therefore has a higher energy content. Consequently, a higher exotherm is observed during its hydrogenation. In general, E-configured alkenes are more stable than the corresponding Z isomers. The situation in cyclooctenes is reversed. As a result of greater ring strain, (E)-cyclooctene has the higher energy content and consequently releases more heat during hydrogenation.

! 79

The structural formula of this new cytostatic (picoplatin) is shown below. In determining the stereodescriptor, the situation arises here that the highest ranking ligand occurs twice. In such cases the priority number of the ligand coordinated *trans* to it with the lowest priority (i.e. with the highest priority number) is used as the configuration index (principle of *trans* maximum difference).

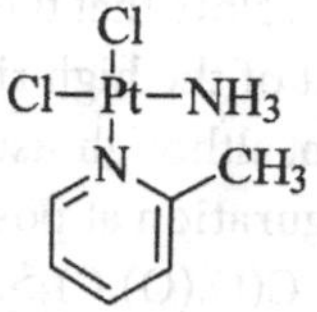

! 80

(*RS,RS*)-2-Phenyl-2-(piperidin-2-yl)acetic acid methyl ester (INN: methylphenidate) is a centrally acting sympathomimetic used in the treatment of attention-deficit/hyperactivity disorder in children. Dexmethylphenidate is the pure (and significantly more active) enantiomer with the *R,R* configuration available in the USA. The diastereomeric racemate with the *RS,SR* configuration whilst a commercially available chemical is not used as a pharmaceutical. (Note the small structural difference between these compounds and levofacetoperane, question 149).

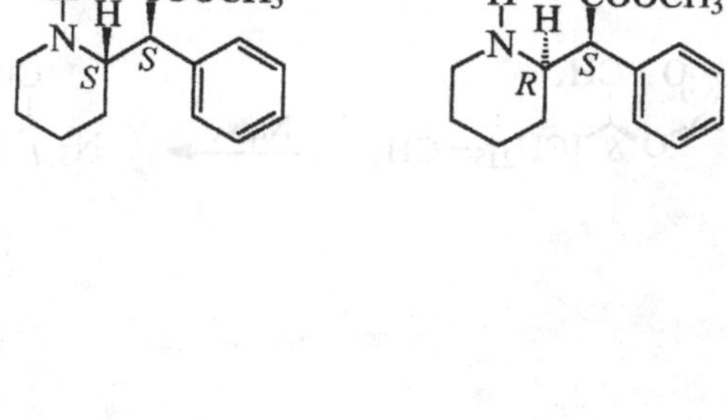

ⓘ 81

The absolute configuration of tazobactam is given in the formula shown below. At position 2 the priority order is N > C(S,C,C) > C(O,O,(O)) > H. The hydrogen atom is directed away from the observer and the S configuration results. At position 3 the priority order is S > C(N,C,H) > C(N,H,H) > C(H,H,H). The priority order at position 5 can be determined in the first sphere: S > N > C > H. The hydrogen atom is directed towards the observer, thus the R configuration results. Note that the nitrogen atom is also a chirality centre since it is, on account of the high ring strain, fixed in a pyramidal arrangement. Its configuration, although usually not given since it is a direct consequence of the configuration at position 5, is S. It is derived from the priority order C(S,C,H) > C(O,(O),C) > C(C,C,H) > e⁻. The unshared electron pair in the formula shown is directed towards the observer.

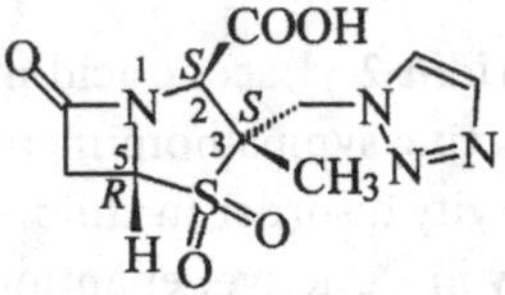

ⓘ 82

The reaction of (S)-1-methylheptyl tosylate with sodium azide is an S_N2 process and affords (R)-2-azidooctane due to inversion. An enantiomeric excess of ee = 99 % indicates that the product is contaminated with 0.5 % of the S isomer. This could theoretically arise as a consequence of a small degree of racemisation during the reaction, due to the reaction having some S_N1 character, although this is improbable. The most likely reason is that the original tosylate had only an ee of 99 % (it is readily prepared from octan-2-ol which is available with an ee of 99 %).

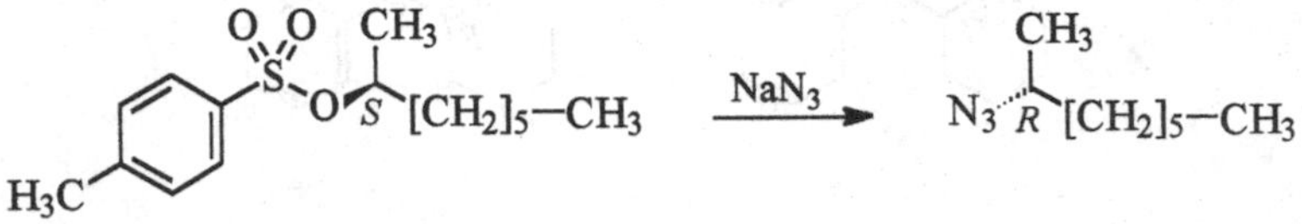

❶ 83

Trimoprostil contains six stereogenic units. The double bonds in positions 5 and 13 have Z and E configuration, respectively. At the chirality centre at position 8 the priority order of the groups is C(O,(O),C) > C(C,C,H) > C(C,H,H) > H and, since the hydrogen atom is directed towards the observer, it has the R configuration. At position 11 it is only necessary to deduce the number of carbon atoms attached to each of the neighbouring carbon atoms to determine the configuration and thus it has also the R configuration. Position 12 likewise has the R configuration. In the first and second spheres no decision can be made since all the neighbouring carbon atoms (C8, C11 and C13) are attached to two carbon atoms, since at C13 a duplicate representation of the carbon atom C14 must be considered as a consequence of the double bond. In the third sphere a decision can, however, be made due to the priority order C9(O,(O),C) > C14(C,(C),H) > C10(C,H,H). The carbon atom at position 15 has the R configuration, since after the hydroxy group (which has the highest priority), the quaternary carbon atom C16 has precedence over the tertiary carbon atom C14. The hydrogen atom (lowest priority) is directed towards the observer.

84

Trovafloxacin contains two mirror symmetric bicyclic systems which can be so arranged that both planes of symmetry coincide. The compound is therefore achiral. The saturated ring system contains three stereogenic centres. The absolute configuration at the two chirality centres C1 and C5 of the bicyclic ring system can be determined by application of the CIP rules. The *R*-configured carbon atom is C1 since it takes precedence over the *S*-configured centre, which is hence C5. A pseudochirality centre is present at C6, which is the reason for the existence of two achiral diastereomers. The amino group has highest priority, followed by the *R*-configured branch and then by the *S*-configured branch. The pseudochirality centre is therefore *s*-configured.

85

a) The reaction of bicyclo[2.2.2]octene with a peracid leads in the first instance to an epoxide which undergoes subsequent hydrolysis in the aqueous acidic solution. Attack by water proceeds by an S_N2 mechanism. Since the original epoxide possesses chirality centres but is achiral, attack by a water molecule can proceed with equal probability at either chirality centre with the result that two enantiomeric *trans*-configured diols are produced.

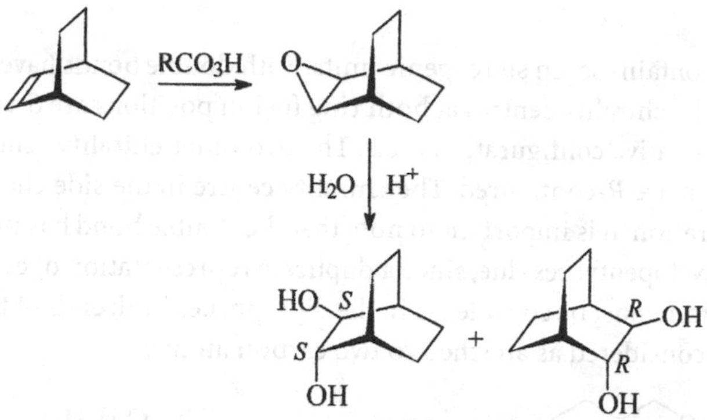

b) Treatment of bicyclo[2.2.2]octene with dilute aqueous potassium permanganate solution initially affords, with concomitant reduction of the manganese, a cyclic manganic(V) ester, which is then hydrolysed to the *meso* product, i.e. the *cis*-diol. Note that attack of permanganate from the other side of the double bond leads to exactly the same product since both bridgehead atoms are not chirality centres.

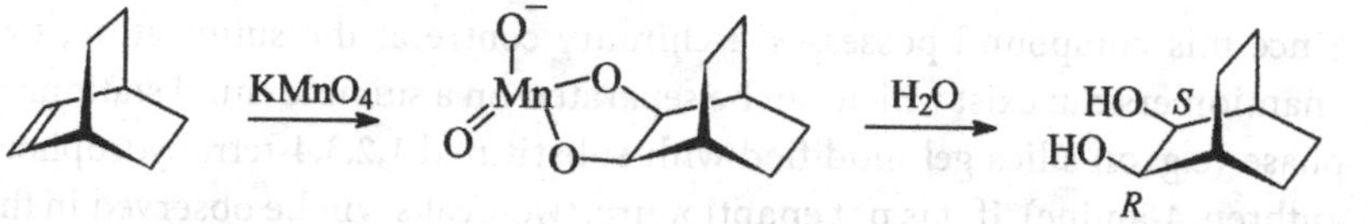

🔴 86

(S,S)-Tartaric acid has a single symmetry element, a C_2 axis, and therefore belongs to the symmetry point group C_2. On rotation through 180° the pairs of carbon centres 1 and 4, and 2 and 3 are transformed into each other. This is also the case if the molecule adopts another conformation as illustrated by the two examples shown below.

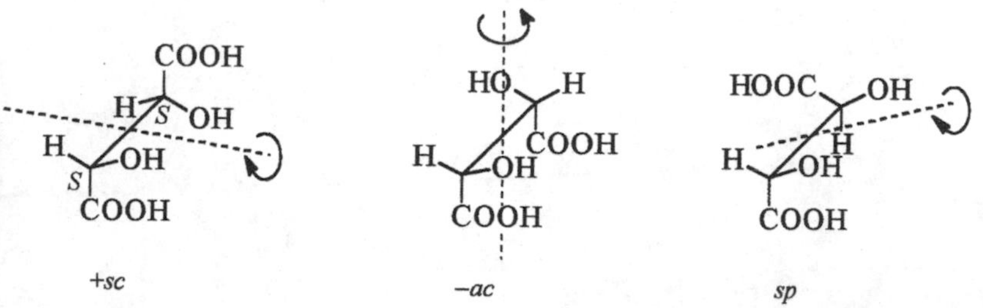

+sc   −ac   sp

❗ 87

Ataprost contains seven stereogenic units. Both double bonds have E configuration. The chirality centres at both ring fusion positions are S-configured and their relative configuration is *cis*. The two other chirality centres in the ring system are R-configured. The chirality centre in the side chain has the S configuration. It is important to note that the double bond has precedence over the cyclopentyl residue, since a duplicate representation of each doubly bonded atom must be considered in the digraph, i.e., both ends of the double bond are considered as attached to two carbon atoms.

❗ 88

Since this compound possesses a chirality centre at the sulfur atom, two enantiomers can exist which can be separated on a suitable chiral stationary phase (e.g. on silica gel modified with substituted 1,2,3,4-tetrahydrophenanthren-4-amine). If it is not enantiopure, two peaks will be observed in the chromatogram.

❗ 89

Maxacalcitol contains nine stereogenic units. These are six chirality centres and two double bonds whose configuration is given in the formula shown below. There is also the single bond between the E and Z double bonds which has partial double bond character. In the formula shown, the conformation of the double bond system is described as s-trans.

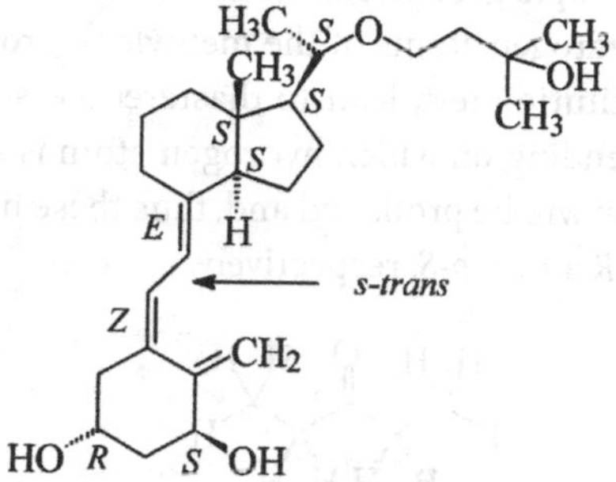

❗ 90

In (2S,3R)-2,3-dichlorocyclobutanone both chlorine atoms are cis. The Re face of the carbonyl group is the less sterically crowded side trans to the chlorine atoms. Attack from this side by hydride ion yields (1S,2S,3R)-2,3-dichlorocyclobutan-1-ol in which the hydroxy group is also cis to the chlorine atoms.

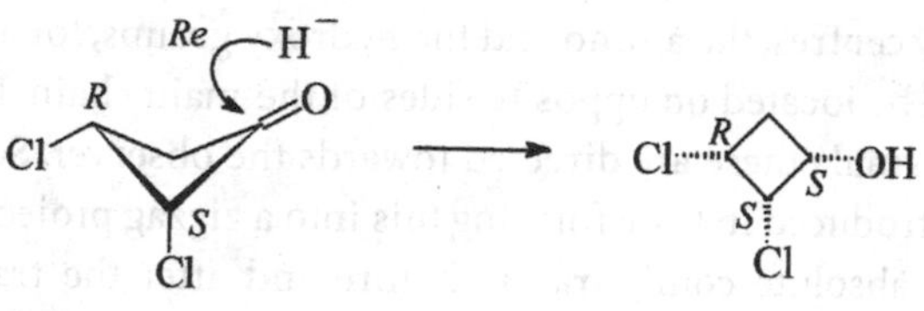

🔵 91

(R)-2-Bromopentan-3-one is a chiral compound which, in addition, also contains two prochirality centres at carbon atoms 3 and 4. All three hydrogen atoms of each methyl group are homotopic, since substitution of one hydrogen atom by another group, e.g. with deuterium, does not produce a chirality centre. It is also of no consequence which hydrogen atom is substituted, since homotopic groups can be transformed into one another by rotation. The two hydrogen atoms of the methylene group would, by applying the above substitution test, lead to diastereomers, therefore these are diastereotopic. Depending on which hydrogen atom is replaced either an R or an S configuration will be produced and, thus these hydrogen atoms have the descriptors pro-R and pro-S, respectively.

🔵 92

In order to derive the structure of the *threo* forms, it is best to start with a Fischer projection formula which should be drawn with the carbon chain vertically oriented and with carbon atom 1 at the top. The substituents at both chirality centres, the amino and the hydroxy groups, for a *threo* configuration must be located on opposite sides of the main chain. In the Fischer projection formula these are directed towards the observer. So, to avoid errors being introduced in transforming this into a zigzag projection, it is best to check the absolute configuration before and after the transformation. Since the descriptor *threo* does not infer the absolute configuration, two enantiomers with the R,R and S,S configuration are possible. Cathine is the S,S isomer. It is important to note that rotation around a single bond does not alter the absolute configuration.

Ph
H—$\overset{S}{|}$—OH
H$_2$N—$\overset{S}{|}$—H
CH$_3$

Ph
H—$\overset{S}{|}$—OH
H$_2$N—$\overset{S}{|}$—H
CH$_3$

Ph
HO—$\overset{R}{|}$—H
H—$\overset{R}{|}$—NH$_2$
CH$_3$

Ph
HO—$\overset{R}{|}$—H
H—$\overset{R}{|}$—NH$_2$
CH$_3$

‖

‖

OH

$\overset{}{\underset{S}{\cdots}}\overset{S}{}NH_2$
CH$_3$

cathine

OH

$\overset{}{\underset{R}{}}CH_3$
$\overset{R}{}$
NH$_2$

❗ 93

The R_a rotamer of afloqualone can be best represented as a projection corresponding to a view along the chirality axis. If the observer looks through the heterocycle for example, along the nitrogen-phenyl bond, then the fluoromethyl residue appears up and the carbonyl group appears down. The methyl group and the hydrogen atom in the phenyl ring can now be considered. Since the priority order is CO > C(N)CH$_2$F > CH$_3$ > H, then in order to obtain the R_a configuration, the methyl group must be located on the right hand side.

H$_2$N— (ring) —N═
                  CH$_2$F
              N
          H
      O H$_3$C

≡

2 CH$_2$F
4 H— ⊖ —CH$_3$ 3
     1 O R_a

❗ 94

In the crossed aldol reaction between acetaldehyde and propiophenone, two chirality centres are created and consequently, four stereoisomers will be produced. Compounds **A** and **B** are enantiomers of each other and can be described with the stereodescriptor *u*. Similarly, **C** and **D** are enantiomers and are *l*-configured. Since both starting materials are achiral, without the use of a chiral base or chiral auxiliary, racemates will be produced. Likewise the choice of base, the addition of a Lewis acid and the reaction conditions used to form the enolate can control which diastereomer is preferentially formed. If the *Z* enolate is formed, the *u* product is the preferred product, whilst the *E* enolate yields predominately the *l* product.

❗ 95

The hydrogen atoms (some of which have been added to the formulae shown in the question) at the prochirality centres in the following formulae have been labelled *pro-R* and *pro-S*. These can be easily confirmed by firstly imagining that the *pro-R* hydrogen atom has a slightly higher priority than the other hydrogen atom and secondly, by determining the configuration of this hypothetical chirality centre. The result would be that the centre is *R*-configured. Note that the central carbon atom in example d) is not a prochirality centre. It could however, be considered as a propseudochirality centre, since substitution of a hydrogen atom leads to a pseudochirality centre. Consequently, the hydrogen atoms can be distinguished as *pro-r* and *pro-s*.

a)

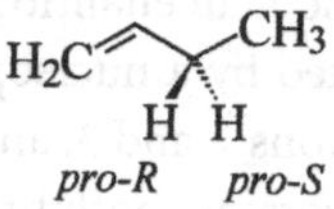

b)

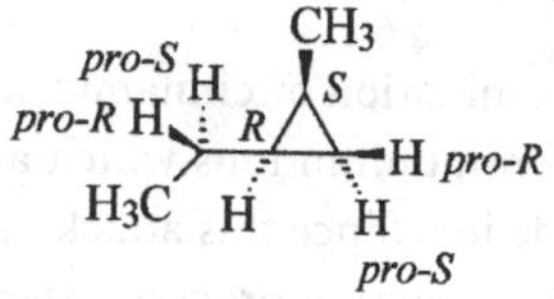

c)

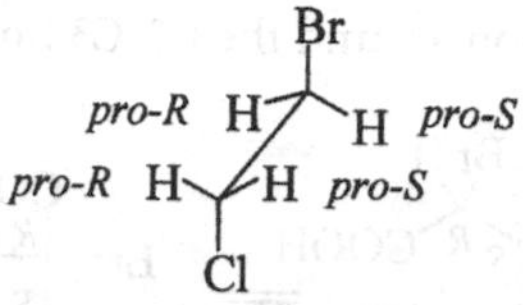

d)

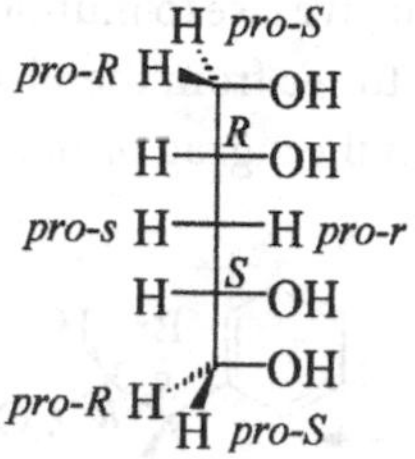

e)

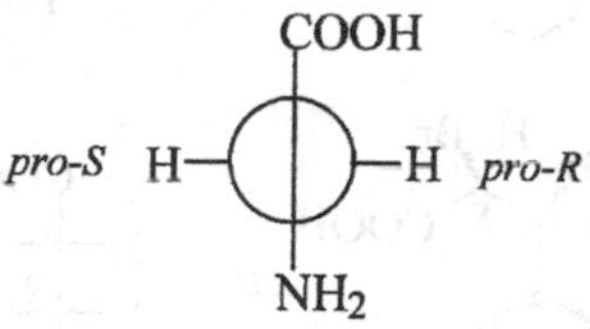

ⓘ 96

The *N*-oxide of loperamide is an achiral compound. If a plane of symmetry is passed through positions 1 and 4 of the piperidine ring, the same groups now appear on each side. Since there are four substituents attached to the piperidine ring, then according to the IUPAC rules the descriptors *cis* and *trans* cannot be used in the name of the compound. They can only be used for rings with two substituents or in general formulation – as in the question – with a precise reference to describe the relative orientation of two defined groups. Here the descriptors *r*, *c* and *t* should be used in the name. (It should, however, be pointed out that *Chemical Abstracts* does not employ the descriptors *r*, *c* and *t*, but uses *cis* and *trans* instead. For their index *cis* and *trans* are defined to refer to the higher ranking groups according to the CIP rules.)

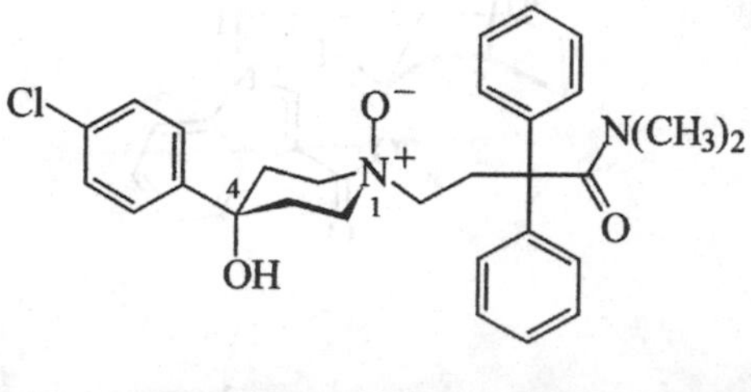

❗ 97

The bromination of cinnamic acid leads to two intermediate enantiomeric cyclic bromonium ions which are subsequently attacked by a nucleophilic bromide ion. Since this attack can occur both at positions 2 and 3, and because the reaction proceeds stereospecifically with inversion, both bromonium ions will give rise to mutually identical pairs of enantiomeric products. This is easier to see from the Fischer projection formulae, which are readily obtained from the zigzag projections by rotation around the C2-C3 bond.

❗ 98

Since the aliphatic ring system of tropisetron is mirror symmetric, both bridgehead atoms (positions 1 and 5) have opposite configuration. Thus position 3 is a pseudochirality centre which is attached to two enantiomorphic residues. Of these enantiomorphic residues, the R-configured branch has priority over the S-configured branch. Therefore C3 has the r configuration. The reader may test this with the mirror image molecule. The pseudochirality centre is still r-configured and thus fulfils the conditions for pseudochirality.

! 99

Darodipine is prochiral. Replacement of a hydrogen atom on one of the methyl groups by a deuterium atom would result in enantiomers, i.e., the methyl groups are enantiotopic. The ester ethyl groups are also enantiotopic. Note that in determining the topicity and the isomeric relationships of the resulting compounds, the wedged bond in the formula could also be drawn as a hatched wedged bond.

! 100

The addition of bromine to (R)-4-chlorocyclohex-1-ene yields initially two diastereomeric cyclic bromonium ions, which then undergo a ring opening reaction with bromide ion. The attack of bromide ion is an S_N2 process and the reaction is stereospecific with inversion. Consequently, the bromo substituents in both diastereomeric products will have a *trans* relationship.

❗ 101

Dizocilpine is a chiral compound with two chirality centres. The methylated centre has the S configuration, whilst the other bridgehead atom is R-configured. For clarification it is helpful to add hydrogen atoms or wedged bonds. Two other possible formulae for the structure are shown below, the last one of which, however, although frequently used is not recommended.

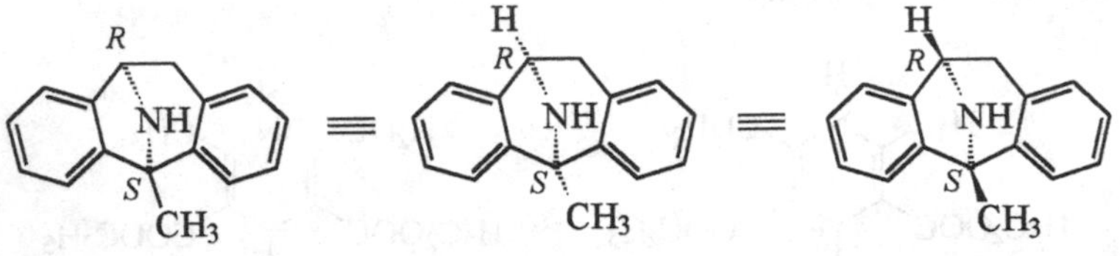

❗ 102

Four stereoisomers can be produced in the reaction. All reaction products have chirality centres at C2 and C6, the absolute configuration of which can be specified by R and S. Compounds **A** and **B** additionally possess a pseudochirality centre at C1 since there are two constitutionally identical but enantiomorphic groups attached to this position. By applying the CIP rules, C1 in **A** has the stereodescriptor r, whilst C1 in **B** has the stereodescriptor s. The groups attached to C1 in **C** and **D** are homomorphic, therefore C1 in these compounds is a prochirality centre.

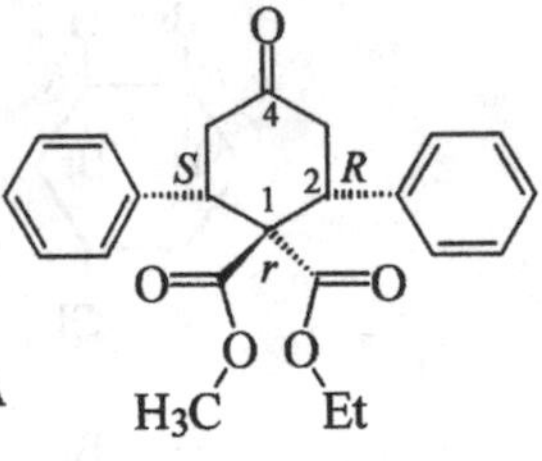

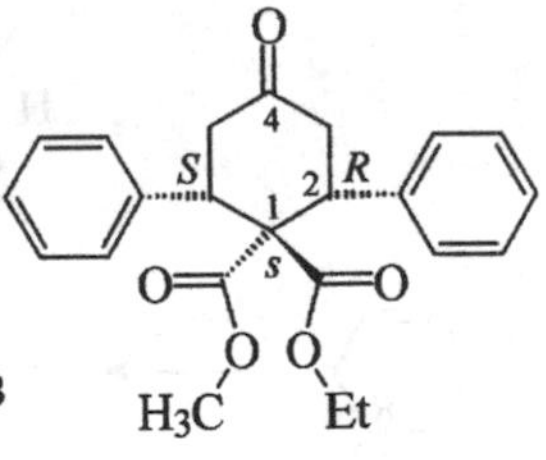

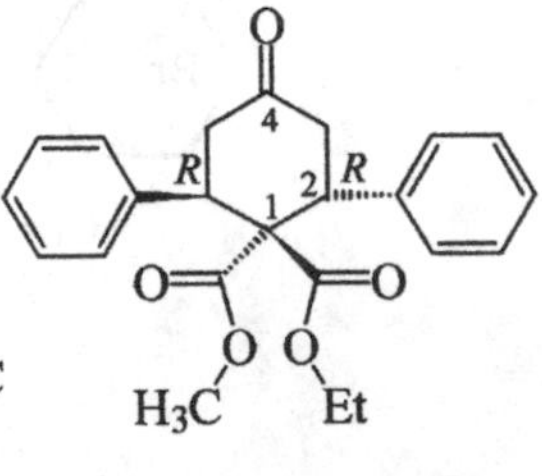

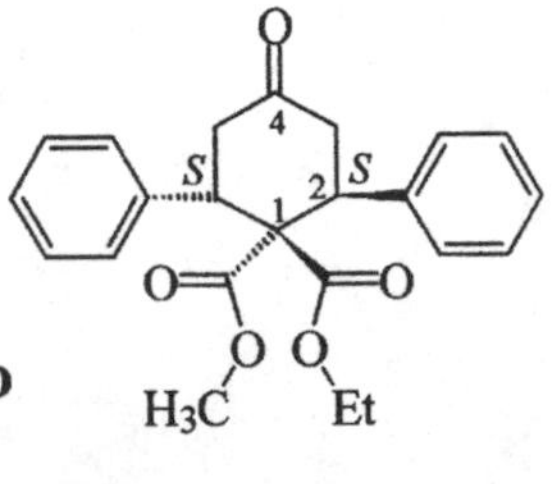

! 103

Sulopenem contains six chirality centres (at one of the sulfur atoms, at the nitrogen atom and at four of the carbon atoms). The hydroxyethyl group has the R configuration. The carbon atom to which it is attached has the S configuration since the priority order of the groups attached to it is C(S,N,H) > C(O,(O),C) > C(O,C,H) > H. At the adjacent carbon atom the priority order is S > N > C > H and this centre is R-configured since the atom of lowest priority is directed towards the observer. The thiolane ring has an S-configured carbon atom and an R-configured sulfoxide. (Note the difference here between a sulfinyl and a sulfonyl group.) Finally, the S configuration can be assigned to the nitrogen atom since the priority order is C(S,C,H) > C(O,(O),C) > C(C,C,(C)) > e⁻, and the unshared electron pair in the formula shown is directed towards the observer.

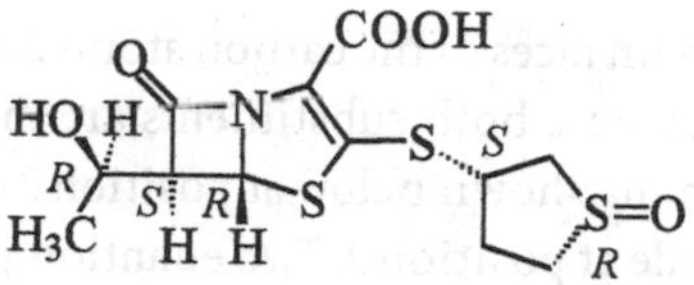

! 104

a) Both faces of the double bond in (E)-1-bromopropene are enantiotopic. At the bromine substituted end of the formula shown below the view is from the Si side and at the other end of the double bond from the Re side.

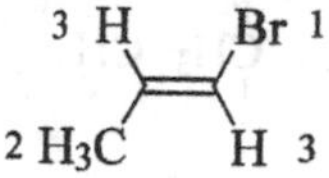

b) At the methyl substituted end of the ethylidene group both faces of the C-C double bond are enantiotopic. The view onto the formula shown below is from the *Re* side. At the other end of the double bond both groups are the same (acetyl groups), and therefore both faces here are homotopic. Both faces of each of the acetyl groups are enantiotopic (the acetyl groups are diastereotopic to one another). In the formula shown below the view is from the *Si* side of the carbonyl group on the right and from the *Re* side of the carbonyl group on the left.

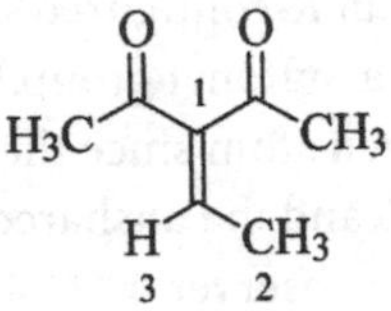

$$\underset{\underset{3}{H}\quad\underset{2}{CH_3}}{\overset{\overset{O\qquad O}{\|\qquad\|}}{H_3C\diagdown\overset{1}{C}\diagup CH_3}}$$

c) In this compound both faces of the carbon atoms 2 and 3 are enantiotopic, whereas at position 4, where both substituents are methyl groups, they are homotopic. In the formula shown below at position 3 the view is from the *Si* side and from the *Re* side at position 2. Two enantiotopic faces formally exist at the nitrogen atom, viewed here from the *Re* side. However, since the trigonal pyramidal structure of hydroxylamines is configurationally unstable, this is in reality only a formal view.

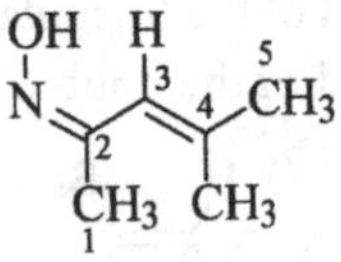

$$\underset{\underset{1}{CH_3}\quad CH_3}{\overset{OH\quad H}{N\diagdown\underset{2}{C}\diagup\overset{3}{C}\diagdown\underset{4}{C}\diagup\overset{5}{CH_3}}}$$

d) The faces of both ends of the double bond are diastereotopic. The hypothetical addition of a group not yet present, e.g. a bromine atom, would produce at the methyl substituted end of the double bond a further chirality centre, whilst at the other end a pseudochirality centre would be formed. In the formula shown below the view at the methyl substituted end is from the *Si* side and at the other end from the *re* side.

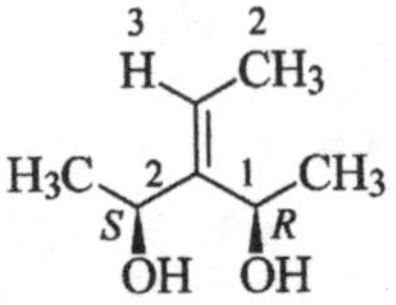

$$\underset{\underset{S}{\uparrow}\quad\underset{R}{\uparrow}}{\underset{OH\quad OH}{\overset{\overset{3}{H}\diagdown\overset{2}{\quad}\diagup\overset{2}{CH_3}}{H_3C\diagdown\underset{2}{C}\diagup\overset{\|}{\underset{1}{C}}\diagdown CH_3}}}$$

! 105

Cefmatilen contains chirality centres at positions 6 and 7 as well as at the β-lactam nitrogen atom. In addition, there is an oxime group. The absolute configuration at C6 is R since the priority order of the atoms directly attached to this chirality centre is S > N > C > H. At C7 the priority order is N > C(S,N,H) > C(O,(O),N) > H and therefore it also has the R configuration. The nitrogen atom has the S configuration since the lone pair of electrons to be considered here is directed towards the observer. The hydroxyimino group is Z-configured, because the higher ranked substituent [C(O,(O),N) > C(C,(C),N)] lies on the same side as the hydroxy group. The configuration of an oxime is relatively stable in contrast to that of a nitrogen atom attached by a single bond.

⊕ 106

Maleic anhydride reacts with cyclopenta-1,3-diene in a Diels-Alder reaction. Since there is a plane of symmetry, the reaction can lead to two achiral compounds, which are diastereomers of each other, containing an *endo*- or *exo*-oriented dicarboxylic anhydride group. These differ in absolute and relative configuration at the bond shared by both rings. Under normal conditions the Diels-Alder reaction proceeds stereospecifically to yield preferentially the *endo* product. Note that in the tricyclic product no *trans* fusion in the ring system is possible as a consequence of the reaction mechanism. Subsequent reduction of the products therefore affords two diols, which are also diastereomers of each other. These may be separated by chromatography on an achiral stationary phase.

⊕ 107

The R_a isomer has the conformation shown in the following formula. The bromo substituent in position 2 of the phenyl ring effectively prevents the two ring systems from being coplanar. The energy required to force the bromo substituted phenyl ring to adopt a torsion angle of 0° has been calculated to be 30.8 kcal/mol. By contrast the equivalent energy required for the compound with an unsubstituted phenyl ring is only 9.1 kcal/mol [2].

$E_{rot} = 30.8$ kcal/mol

🛈 108

The bridgehead atom at position 1 has R configuration. The priority order of the groups attached to this position is N > C(C,C,H) > C(C,H,H) > H. At position 2 the priority order of the attached groups is C(N,(N),H) > C(N,C,H) > C(C,C,H) > H. It is important to note here that the doubly bonded nitrogen atom is because of the duplicate representation regarded as two nitrogen atoms. At C3 the arrangement according to the priority order is in counter-clockwise direction to the observer, and therefore this has the S configuration. The second bridgehead atom is S-configured. Here the lowest priority substituent, the hydrogen atom, can only be directed away from the observer. The configuration of the double bond is E, because the substituents lie on opposite sides of the double bond. As well as the absolute configuration, it is also possible to determine the relative configuration at C2 and C3. The substituent at position 2 is on the opposite side of the main bridge (the nitrogen atom) and therefore is *endo*-oriented whilst the dichlorophenyl ring is *exo*-oriented.

❶ 109

(2R,3s,4S)-2,3,4-Trichloropentanedioic acid is a meso compound. Position 3 is a pseudochirality centre since two enantiomorphic groups are bonded here. The s configuration is assigned, because the R-configured group has priority over the S-configured group in the CIP rules.

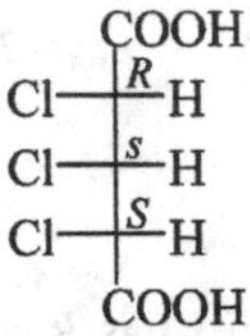

❶ 110

The configuration of all the chirality centres is given in the formula shown below. For the chirality centre at position 7 in the ring, the digraph indicates how the priority order is determined.

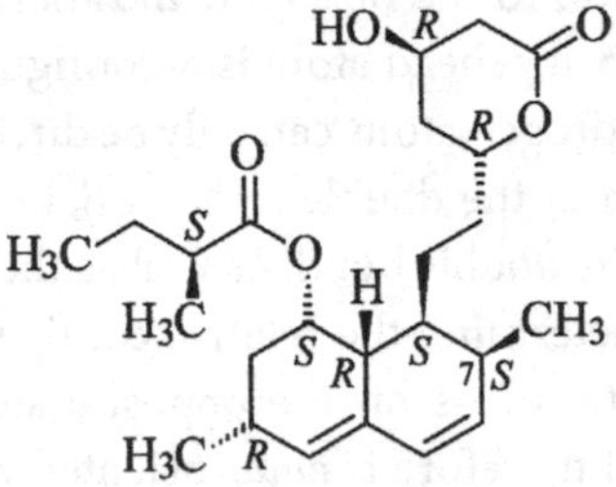

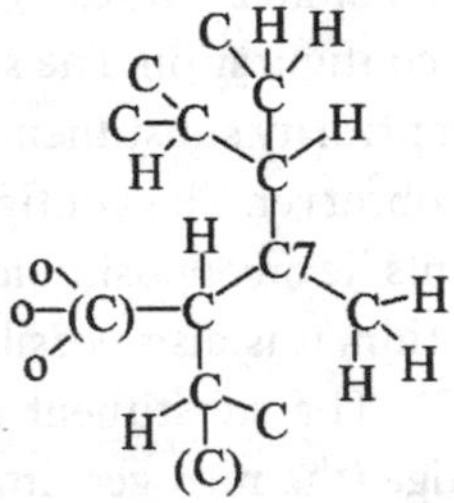

❶ 111

The compound possesses a chirality plane. Before establishing the configuration, it is necessary to determine the pilot atom. This is the atom of highest priority according to the CIP rules outside the chirality plane that is directly bonded to an atom in the plane. This is indicated in the formula below with an arrow. A view from this atom onto the chirality plane shows that the atoms on the way to the cyano group will be reached by a counter-clockwise curve. The configuration is therefore S_p.

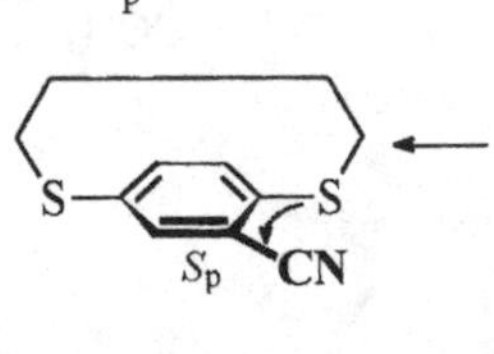

❗ 112

a) The cyclic bromonium ion produced from maleic acid is mirror symmetrical. Since there is equal probability that subsequent nucleophilic attack by methanolate can take place at either carbon atom, the product is a racemate. The reaction is not enantioselective but it is stereoselective since the *u*-configured racemate is formed exclusively as a product because the ring opening reaction proceeds with Walden inversion.

b) The cyclic bromonium ion produced from fumaric acid has only a two-fold proper axis of symmetry. Thus it will exist as a pair of enantiomers. Since there is equal probability that subsequent nucleophilic attack by methanolate can take place at either C2 and C3 of either enantiomer, the same *l*-configured product will be obtained in either case and consequently, overall a racemate will be obtained. The reaction is not enantioselective but is stereoselective. Since maleic acid and fumaric acid are stereoisomers and since the product obtained here is stereoisomeric to that obtained from maleic acid, the reaction is therefore also stereospecific.

❗ 113

For a β-elimination of HBr from (1S,2R)-1-bromo-2-fluoro-1,2-diphenyl-ethane, a base is required to abstract a hydron from the position β to the bromine atom. The reaction has an E2 mechanism. Attack by the base, formation of the double bond and elimination of bromide ion take place in a concerted process. A prerequisite is that both the groups eliminated are arranged antiperiplanar to each other. Consequently, the resulting olefin is E-configured. The Z-configured olefin is the reaction product from both the R,R- or the S,S-configured isomers for the same reason. This is therefore a stereospecific reaction since the absolute configuration of the starting material dictates the configuration in the product.

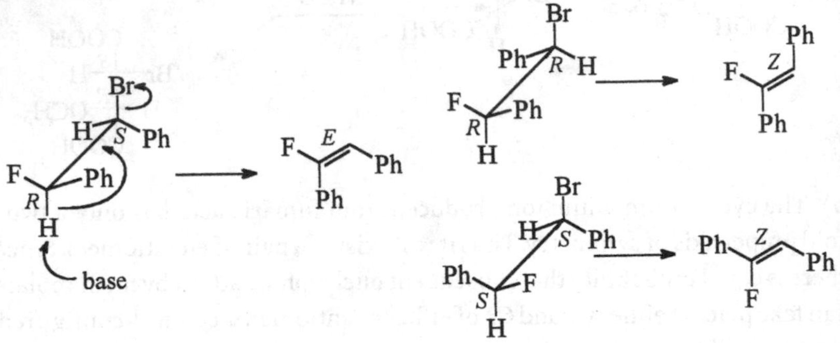

❗ 114

Eplivanserin contains three stereogenic units. In addition to the E-configured alkene double bond there is a Z-configured substituted oxime. The single bond between both double bonds has partial double bond character. The formula shown below represents the s-trans isomer. This means that it represents an antiperiplanar conformation.

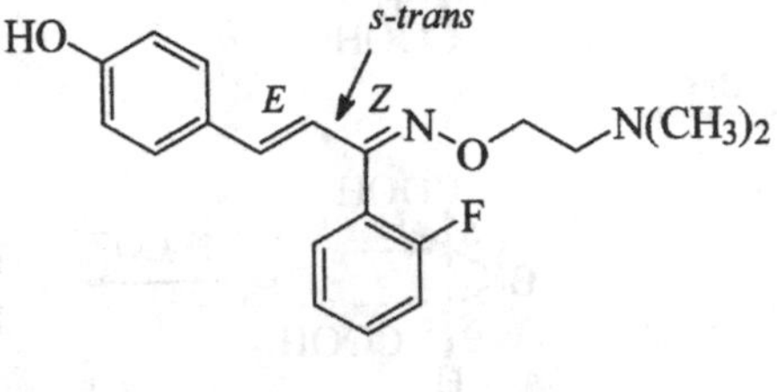

❗ 115

a) In this phosphane the unshared electron pair on phosphorus has the lowest priority. The compound has therefore the S configuration.

b) This phosphinate has also the S configuration. It is important to note here that the contribution of the d-orbitals to the double bonds is, in general, ignored. As a consequence, at the P-O double bond no duplicate representation is considered in the determination of the absolute configuration. Therefore the methoxy group has precedence over the doubly bonded oxygen atom.

c) The lone pair of electrons on the sulfur atom, that is responsible for the pyramidal structure, has the lowest priority of the groups attached to the chirality centre. This is directed away from the observer and therefore the sulfoxide represented by this formula has the S configuration.

d) The absolute configuration of this coordination compound can be described by the stereodescriptor A. To determine this, the observer must look from the ligand of highest priority to the four ligands lying in the plane below it and deduce whether these ligands according to the priority order derived from the CIP rules have a clockwise or counter-clockwise arrangement. Since the order is bromide, pyridine, nitrite and ammonia (pyridine has a higher priority than ammonia), the arrangement viewed from the iodine atom is anticlockwise and therefore the compound has the A configuration.

OC-6-34-A

❗ 116

In order to deduce the structure of the desired diastereomer, the observer must view the molecule from the side and move one of the two benzene rings in a clockwise direction. This leads to a torsion angle at the chirality axis between the two aromatic rings. This twisting is caused by the mutual steric hindrance of the methoxy groups in the two benzene rings and the resistance of the eight membered ring to adopt a planar conformation. Although this is an idealised description, the rotamer can be shown as a projection akin to the Newman projection. From this it is clear that the compound can be viewed from either side. At both ends of the chirality axis there is a methoxy group (priority numbers 1 and 3) which has a higher priority than the methylene groups (priority numbers 2 and 4).

! 117

! 118

Oxilofrine must first be drawn as a Fischer projection formula since the stereodescriptor *erythro* indicates the relative orientation of the heteroatom groups in a Fischer projection formula of the molecule. Because the stereo-descriptor *erythro* only defines a relative configuration, two enantiomers must be considered. Next the Fischer projection formulae are converted into zigzag projections. It is advisable as a check to determine the configuration at the chirality centres in all the formulae, as it cannot change during the change from one projection to the other.

❗ 119

At all three chirality centres the hydrogen atoms must be indicated by a bold wedged bond. The priority order at both ring fusion positions is N > C(S,...,H) > C(N,C,H) > H and at position 4 it is S > C(N,C,H) > C(C,H,H) > H.

❗ 120

In changing from one method of drawing the formula to another, it is important that not only constitution remains unaltered but also that the configuration of all the stereogenic units remains the same. Errors can be avoided by checking the absolute configuration of the chirality centres in the original formula given (in this case the Haworth projection) with those in the required formula.

❗ 121

The chemically equivalent hydrogen atoms are those with the same numbers in the following formulae.

a)

b)

c)

① 122

In the aldol reaction between butanone and benzaldehyde it is important to note that butanone can form more than one enolate. Under the reaction conditions given, formation of the more substituted and thermodynamically more stable enolate will be produced, and this can be either *E*- or *Z*-configured. Moreover, nucleophilic attack at the aldehyde group of the planar benzaldehyde can take place both from the *Re* or *Si* sides. Four products are therefore obtained. From the *Z* enolate the *l*-configured enantiomers **A** and **B** are the preferred products, whilst the *E* enolate gives predominately the *u*-configured enantiomers **C** and **D**.

① 123

The compound possesses a chirality axis that passes through the double bond and carbon atom 1. In order to determine the absolute configuration the observer looks along this axis, for example from C1 where the hydroxy and bromomethyl groups are arranged vertically. The substituents at the methylidene group on C4 are horizontally oriented. The priority order of these four groups can then be determined, which leads (in this example) to an S_a configuration. The compound is therefore (S_a)-1-(bromomethyl)-4-[chloro(methoxy)methylidene]cyclohexanol.

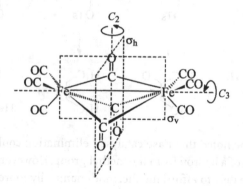

124

The symmetry elements are three C_2 axes each passing through a bridging carbonyl group and the centre of the molecule, three vertical planes of symmetry σ_v each containing both iron atoms and a bridging ligand, and a horizontal plane of symmetry σ_h at right angles to the main C_3 axis of symmetry and contains all the bridging ligands. These results can be verified by using the flow chart in the appendix. [$Fe_2(CO)_9$] therefore belongs to the symmetry point group D_{3h}. It should be noted that the actual structure of this binuclear iron complex in the crystalline state differs marginally from this ideal symmetry [3].

🛈 125

The first step is to convert the Fischer projection formulae into sawhorse projections. Note that for a base-induced elimination reaction of the E2 type, the tosyl group must be antiperiplanar to the proton or deutron being abstracted from the neighbouring carbon atom. As a consequence of this reaction mechanism, the configuration of the resulting double bond can be predicted. Because of the higher steric hindrance in the transition states leading to the Z isomers, in both cases the E isomer should be expected as the major product.

It should also be noted that base catalysed elimination could also occur by the abstraction of a hydron from the methyl group. However, the E2 reaction leads predominately to afford the thermodynamically more stable product, i.e., that with the higher substituted double bond.

🛈 126

The synclinal and anticlinal conformations of 2-chloroethanol are chiral, which are designated as $+sc$ and $-sc$, and $+ac$ and $-ac$ depending upon the direction of rotation. In addition, with the exception of the synperiplanar and antiperiplanar forms, all other conformations are chiral, i.e. there are an infinite number of chiral conformations.

−sc   −ac   +sc   +ac

❶ 127

Cinchonine contains five chirality centres. Note that because of the rigid structure of the molecule, the nitrogen atom in the quinuclidine ring is also a chirality centre and has the S configuration because the group of lowest priority (the unshared electron pair) is directed towards the observer.

❶ 128

Because the amide C-N bond has partial double bond character, there is restricted rotation around this bond. Therefore, in addition to the three signals expected for the thiophene protons in the ^1H NMR spectrum, there are two broad singulets for the two amide protons since these are both chemically and magnetically non-equivalent. The amide hydrogen atoms are diastereotopic and therefore can be distinguished by the descriptors pro-Z and pro-E.

❶ 129

1-Methylcyclopenta-1,3-diene reacts as a diene with maleic anhydride as dienophile in a Diels-Alder reaction to yield four stereoisomers which are diastereomeric pairs of enantiomers. Enantiomers are the two compounds with the *endo*-oriented dicarboxylic anhydride group and the two compounds in which this group has the *exo* orientation.

❶ 130

This compound is isobornyl acetate which has an *exo*-oriented acetate group and therefore the compound will exist as two enantiomers, i.e., the *R,R,R* isomer and its non-superimposable mirror image *S,S,S* isomer. The name given in the question designates the racemate.

❶ 131

Because the stereodescriptor *cis* indicates only a relative configuration of the two chirality centres in the ring, there will be two isomers of *cis*-1-[(R)-*sec*-butyl]-2-methylcycohexane. Since the *sec*-butyl side chain of both isomers must be *R*-configured, the two compounds will be diastereomers. Each diastereomer can exist in two possible chair conformations. The more energetically favourable conformations will be those where the *sec*-butyl group occupies an equatorial position, since in these the more bulkier ethyl group will be at its maximum possible distance from the ring system.

132

The compound has a chirality axis with the S_a configuration. In order to determine the configuration, the four substituents in the two perpendicular planes at the forward and rear quaternary centres are drawn in a projection corresponding to a view along the chirality axis. The groups nearer to the observer have priority over the groups further away. A curve drawn from the closer situated ethyl group to the closer situated methyl group and then towards the other ethyl group will be counter-clockwise.

chirality axis

133

🟡 134

The conversion of pentan-2-ol to pentan-2-amine involves the formal substitution of the hydroxy group by an amino group. For the reaction to proceed in a stereospecific manner, the hydroxy group must first be converted into a better leaving group such as the acetate by esterification with acetic acid or better by preparing the tosylate by treating the alcohol with tosyl chloride. The reaction of the tosylate with sodium azide in an S_N2 reaction affords 2-azidopentane which can then be reduced to the amine by treatment with lithium aluminium hydride or with triphenylphosphane. Since substitution of the tosylate involves inversion, the starting material for the preparation of (S)-pentan-2-amine as outlined above is (R)-pentan-2-ol.

Note that alkyl azides are potentially explosive. This and the simpler reaction conditions involved is one reason why the reaction described above is mainly carried out as a Mitsunobu reaction. In this method the alkanol is reacted with diethyl diazenedicarboxylate, triphenylphosphane and hydrazoic acid in a "one-pot" reaction. The reaction also proceeds with inversion and affords the amine directly because the intermediate azide is reduced by excess triphenylphosphane *in situ*.

🟡 135

The stereodescriptor *erythro* indicates that the chlorine atom and the hydroxy group lie on the same side of the main chain in a Fischer projection formula. Since it does not define an absolute configuration, two enantiomers must be considered. The Fischer projection formula, which represents an eclipsed conformation, is simplest first converted into a sawhorse projection (also in its eclipsed conformation) and then one side of the molecule rotated until both the reference groups, the chlorine atom and the hydroxy group, adopt an antiperiplanar arrangement. The required Newman projection formula can then be derived from these formulae.

erythro   *ap*

! 136

! 137

The epimers of (2R,4aR,8aR)-decahydronaphthalen-2-ol are (2S,4aR,8aR)-decahydronaphthalen-2-ol, (2R,4aS,8aR)-decahydronaphthalen-2-ol and (2R,4aR,8aS)-decahydronaphthalen-2-ol. Note that epimers differ only at one chirality centre in their absolute configuration.

❶ 138

$[Ca(EDTA)]^{2-}$ has only a C_2 axis and no plane of symmetry and consequently belongs to the symmetry point group C_2. The complex is therefore chiral. The enantiomers have the descriptors $OC\text{-}6\text{-}2'1'\text{-}C$ and $OC\text{-}6\text{-}2'1'\text{-}A$. The prime symbols added to some of the priority numbers are used to distinguish between the two nitrogen atoms and the carboxymethyl groups attached to them. Priority numbers without the prime symbol have precedence over the same priority numbers with the prime symbol. In determining the configuration as well as the absolute configuration (denoted by the chirality symbols A and C), it is irrelevant which nitrogen atom becomes primed.

OC-6-2'1'-C   OC-6-2'1'-A

❶ 139

The stereodescriptor *sn* indicates that the name results from the convention of stereospecific numbering of the atoms of glycerol. The structure must therefore be drawn as a Fischer projection formula with the L configuration.

🟠 140

The structure of omapatrilat contains an L-homocysteine residue, which is highlighted in the formula shown below. In a published synthesis of this compound, the first step is the separation of enantiomers of a homocysteine derivative [4].

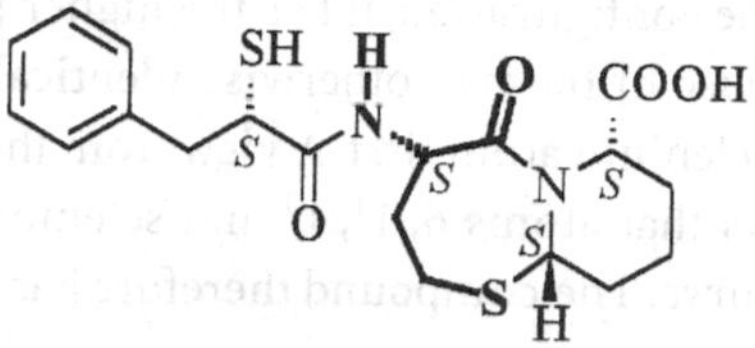

🟠 141

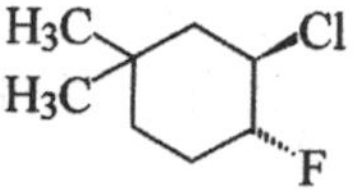

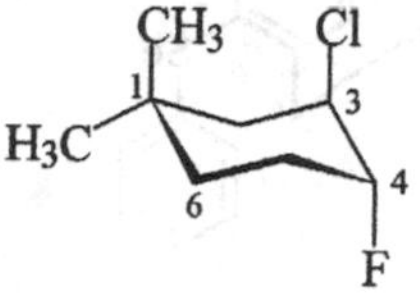

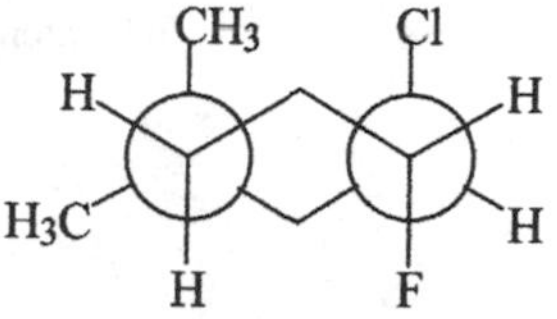

🟠 142

The stereodescriptor for the anion of this compound is *TBPY*-5-11 and placed in round brackets (and in this example – in the absence of the cation – at the beginning of the name): (*TBPY*-5-11)-Bis(*tert*-butyldiphenylsilyl)-tris(dimethylamido)hafnate(1–). The polyhedral symbol *TBPY*-5 indicates a trigonal bipyramidal structure with the coordination number 5 and the configuration index specifies the priority numbers (in ascending order), determined according to the CIP rules, of the two apical ligands.

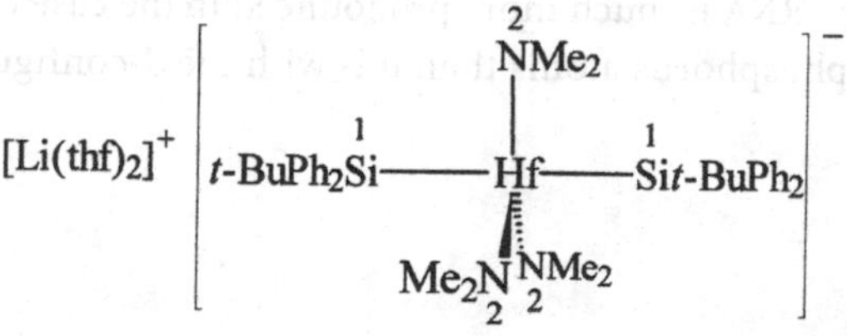

⊕ 143

The compound contains an *S*-configured chirality centre in the 4,5-dihydro-1,3-oxazole ring as well as a chirality plane, i.e., the benzene ring with the locant 1 in the cyclohexaphane skeleton. The pilot atom for determining the configuration is atom 5, since of the two atoms 3 and 5, which might be used to determine the configuration, it has the higher priority according to the CIP system because of the two otherwise identical pathways that from atom 5 reaches the selenium atom first. A view from the pilot atom onto the chirality plane shows that atoms 6, 1^4, 1^3 and selenium are arranged on a counter-clockwise curve. The compound therefore has the S_p configuration.

⊕ 144

Although the reaction proceeds with retention of configuration, the descriptor for the configuration at the phosphorus atom changes since there is a new priority order for the substituents attached to it. Note that the ethoxy group attached to the phosphorus atom has a higher priority than the doubly bonded oxygen atom because the P-O double bond is not considered and accordingly, there is no duplicate representation of atoms to be added. (It has been shown from the incorporation of such compounds in oligonucleotides, that the stabilising effect of such fragments in DNA-duplexes with complementary RNA is much more pronounced in the case of epimers with *R*-configured phosphorus atoms than it is with the *S*-configured diastereomers [5].)

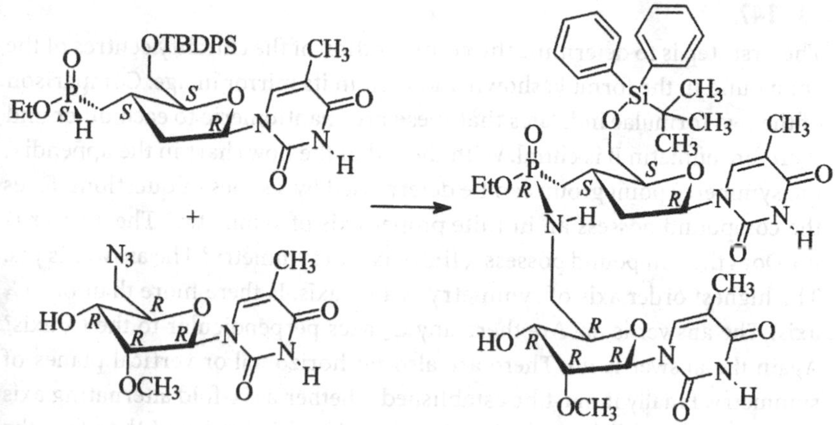

145

The compound has the *R* configuration. Note that the unshared electron pair on the sulfur atom in the formula shown below is oriented towards the observer. Contributions of the sulfur d-orbitals to the S-O double bond are not considered in the determination of the priority order of the groups attached to the chirality centre. Thus the decyloxy residue has precedence over the doubly bonded oxygen atom. This compound has been used in the synthesis of liquid crystalline substances [6].

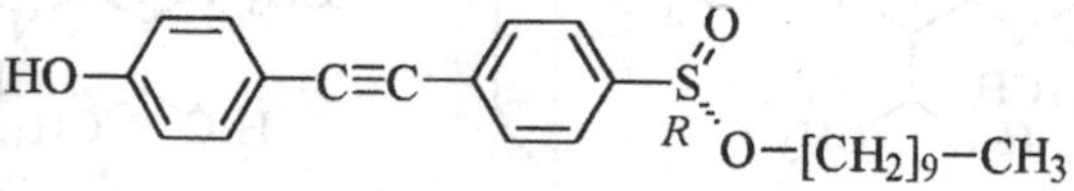

146

As can be seen in the formulae of both *endo*-substituted enantiomers of renzapride shown below, if only the relative configuration is considered then the formulae are reflection invariant, whilst the absolute configuration is inverted by a mirror reflection.

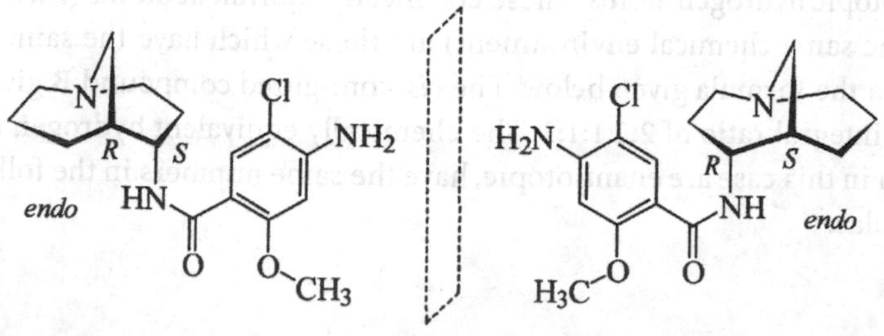

❗ 147

The first step is to determine the configuration of the chirality centres of the compound in the formula shown and then in its mirror image. Comparison of the two formulae indicates that these are enantiomeric to each other and therefore enniatin B is chiral. With the aid of the flow chart in the appendix, the symmetry point group can be determined by a series of questions. Does the compound possess an infinite proper axis of symmetry? The answer is no. Does the compound possess a finite axis of symmetry? The answer is yes. The highest order axis of symmetry is a C_3 axis. Is there more than one C_3 axis? The answer is no. Are there any C_2 axes perpendicular to the C_3 axis? Again the answer is no. There are also no horizontal or vertical planes of symmetry. Finally it must be established whether a six-fold alternating axis of symmetry parallel to the C_3 axis exists. There is none and therefore the symmetry point group to which enniatin B belongs is C_3.

enniatin B   *ent*-enniatin B

❗ 148

trans-1,3-Dichlorocyclopentane (A and/or *ent*-A) has signals with an intensity ratio 1:1:1:1 in its ^1H NMR spectrum, because there are four pairs of homotopic hydrogen atoms. These chemically equivalent atoms (each group has the same chemical environment) are those which have the same numbers in the formula given below. The *cis*-configured compound B gives rise to an integral ratio of 2:2:1:1:2. The chemically equivalent hydrogen atoms, which in this case are enantiotopic, have the same numbers in the following formulae.

A 1:1:1:1   *ent*-**A** 1:1:1:1   **B** 2:2:1:1:2

❗ 149

The first step is to expand the zigzag projection given by including the groups directed towards the observer and to determine the configuration of the chirality centres. This must not alter in the subsequent transformation. Since the formula already represents the same conformation as the required Fischer projection formula, the perspective representation can be transformed into the Fischer projection formula in only one intermediate step. It should be noted that the stereodescriptors *threo* and *erythro* only indicate the relative configuration of the heteroatom groups in the Fischer projection formula. Levofacetoperane is *threo*-configured, since in this case they lie on opposite sides of the vertically oriented carbon chain. (Note the slight structural difference between this compound and methylphenidate in question 80.)

zigzag projection

threo

Fischer projection formula

150

❶ 151

The compound is an amide and the amide C-N bond due to partial double bond character has restricted rotation. The signals of the resulting *E* and *Z* isomers which are, on the NMR time scale, sufficiently long lived are therefore distinguishable in the spectrum.

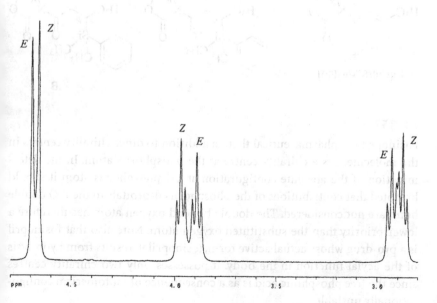

❗ 152

Lithium cuprates react with α,β-unsaturated carbonyl compounds by 1,4-addition. The diastereomers **A** and **B** will be the expected products whose relative proportions will depend upon the precise reaction conditions. On the assumption that stabilisation of the starting material in the conformation shown occurs through complexation with metal ions a preferential attack of the silyl group will take place from the side nearest the reader, i.e. from the least hindered side, and hence a preponderance of **A** in the product mixture will be expected. It was actually observed that not only the yield but also the diastereomeric ratio was strongly dependent upon the amount of dimethylsulfane in the reaction mixture (or complexed to the cuprate). In the presence of 0.75 mol% dimethylsulfane with respect to the added cuprate, the ratio of **A** to **B** is dramatically increased (diastereomeric ratio 97 : 3), whilst in the absence of dimethylsulfane it is approximately 1 : 1 [7].

❗ 153

Fosinopril is a pharmaceutical that, in addition to other chirality centres in the molecule, has a chirality centre at the phosphorus atom. In the determination of the absolute configuration at the phosphorus atom it should be noted that contributions of the phosphorus d-orbitals to the P-O double bond are not considered. The doubly bonded oxygen atom has therefore a lower priority than the substituted oxygen atom. Note also that fosinopril is a pro-drug whose actual active form, fosinoprilat, results from hydrolysis of the acylal function in the body. It possesses only two chirality centres since the free phosphinic acid is as a consequence of tautomerism configurationally unstable.

fosinopril

fosinoprilat

🛑 154

There are three vicinal dichlorinated stereoisomers of dichlorocyclopropane and one geminal dichlorinated isomer **A** (with a two-fold axis of symmetry and two vertical planes of symmetry containing this axis; symmetry point group C_{2v}). Since all the hydrogen atoms in the molecule can be transformed into one another by symmetry operations, all are equivalent and the ^{1}H NMR spectrum contains only one signal. The *cis*-configured compound **B** is achiral and is a meso compound (with a plane of symmetry; symmetry point group C_s). Those hydrogen atoms which are reflections of one another in the plane of symmetry are equivalent and yield a signal of double intensity. Therefore a spectrum with the intensity ratio of 2:1:1 will be expected. **C** and **D** are *trans*-configured, chiral and consequently enantiomers of each other. Each of these isomers only possesses a two-fold axis of symmetry and belongs to the symmetry point group C_2. The hydrogen atoms will appear at two positions in the spectrum, since those hydrogen atoms which can be transformed into each other by a C_2 operation will be equivalent. Chemically equivalent hydrogen atoms are those with the same numbers in the formulae.

A   **B** 2:1:1   **C** 1:1   **D** 1:1

● 155

a) The compound is unambiguously characterised by indicating the configuration of the pseudochirality centre. In order to determine this configuration, however, it is first necessary to determine the configuration of the chirality centres. Since the compound is achiral, it is also unambiguously characterised, if one specifies that the indol-3-yl substituent occupies an *exo* position. Two diastereomers are possible, which can be distinguished between from the position of this group and consequently by the configuration of the pseudochirality centre.

b) In this compound *E/Z* isomerism at the double bond is possible. The *R*-configured branch of the bicyclic ring system has precedence over the *S*-configured branch. The two possible isomers are enantiomers.

c) The compound contains two chirality centres. Therefore enantiomerism and diastereomerism are possible (there is a total of three stereoisomers because a meso compound exists). Note that in determining the configuration the *tert*-butyl group is the lowest ranked group. The methylene group, since it is attached to a silicon atom, has precedence over the phenyl group.

d) The compound is unambiguously characterised by indicating the configuration of the double bond between the two ring systems. Since there are two enantiomorphic groups attached to one end, the *E/Z* isomers are enantiomers.

e) There is an enantiomer and diastereomers of this compound. The enantiomer is the *E* isomer.

❗ 156

After determining the configuration of the chloroethyl groups, the symmetry point group can be deduced with the aid of the flow chart given in the appendix. Since the molecule is non-linear and therefore cannot have an infinite axis of symmetry, the n-fold axis of symmetry of the highest order must be determined. This is a single C_2 axis. It is therefore unnecessary to pose the question whether this is perpendicular to other C_2 axes. Neither a horizontal plane of symmetry nor a vertical plane of symmetry exist, therefore it is now only necessary to find out whether the C_2 axis is also an S_4 axis. This is the case because a rotation through 90° followed by reflection at a perpendicular plane produces the same molecule. The symmetry point group is therefore S_4 with the two symmetry elements C_2 and S_4.

C_2, S_4

❗ 157

The compound has an S-configured chirality centre in the side chain. In addition, there are two chirality planes as both benzene rings are unsymmetrically substituted. The pilot atom for the determination of the configuration must be deduced separately for each of the chirality planes. For the chirality plane closer to the observer (the bromo-substituted ring) it is atom 2, whilst for the other chirality plane it is atom 3. Thus for the benzene ring with the locant 1 the S_p configuration results and for the other benzene ring the R_p configuration. This compound has been synthesised enantiomerically pure [8].

● **158**

The compound has a total of eight prochirality centres, one at every carbon atom with the exception of the methyl groups. Of special significance for the topicity of the hydrogen atoms is the prochirality centre at position 2 in the ring. In the side chain each of the hydrogen atoms at both methylene groups are enantiotopic. By contrast, the hydrogen atoms at both the ring methylene groups are diastereotopic. This is because the methylene groups are themselves already enantiotopic groups of the prochirality centre at position 2 which will become a chirality centre if a hydrogen atom in the ring is substituted. It is especially worth noting that for each of the ring hydrogen atoms another enantiotopic hydrogen atom exists, i.e., the hydrogen on the neighbouring methylene group that is *cis* to it.

● **159**

According to Cram's rule, the *u*-configured pair of enantiomers (the 2*R*,3*S* and 2*S*,3*R* isomers) will be the preferred products when *rac*-2-methylbutanal reacts with hydrogen cyanide. The byproduct will be the *l*-configured racemate (2*RS*,3*RS*)-2-hydroxy-3-methylpentanenitrile.

In order to determine the product distribution by Cram's rule, the structural formula is drawn such that the oxygen atom is antiperiplanar to the largest group of the neighbouring chirality centre, the ethyl group in this example. The nucleophile (CN^-) then attacks from the less hindered side. This is shown below for one of the enantiomers.

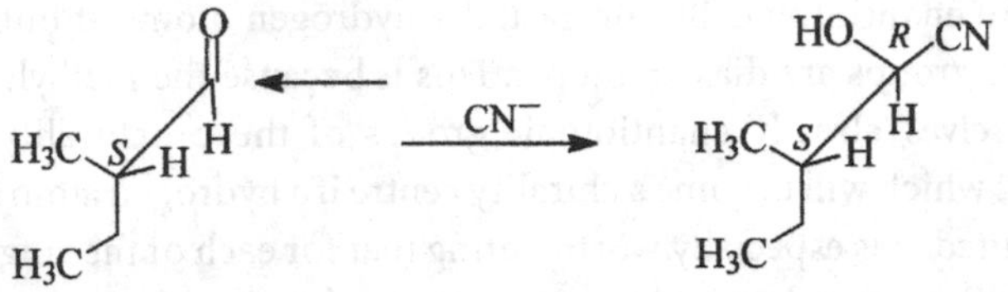

⊕ 160

The observation that all neighbouring *sec*-butyl groups are mutually *trans* to one another with alternating R and S configuration, simplifies the determination of the symmetry point group using the flow chart given in the appendix. Since the molecule is non-linear, there is no infinite axis of symmetry. The n-fold axis of symmetry of the highest order present is a C_3 axis. There are no C_2 axes perpendicular to the C_3 axis, neither is there a horizontal plane of symmetry nor a vertical plane of symmetry. The only remaining question is whether the C_3 axis is also an S_6 axis. This is the case because a rotation through 60° followed by reflection at a perpendicular plane produces the same molecule. The symmetry point group is therefore S_6. In addition to the symmetry elements C_3 and S_6, the molecule also contains a centre of symmetry.

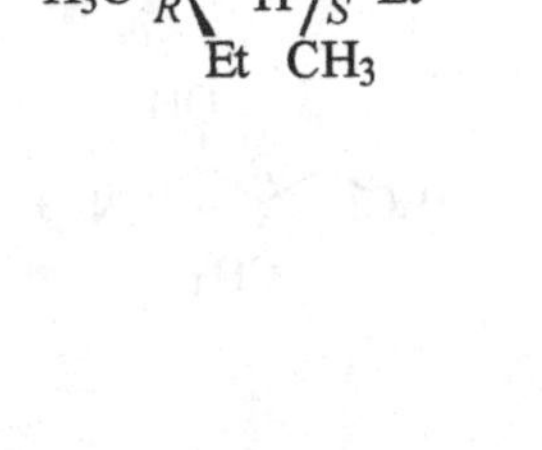

C_3, S_6

❶ 161

The base will abstract a proton from one of the methylene groups α to the carbonyl group. These methylene groups are enantiotopic groups of the prochirality centre at the spiro atom. The *pro-R* group is that which during the course of the reaction will actually become the higher ranking. The *R* product is therefore obtained. Since enolate formation with a strong base at low temperature is irreversible, if 92 % of base attack occurs at the *pro-R* group then 92 % of the *R* product will be produced. In the reaction mixture there will hence be 8 % of the *S* enantiomer. The amount of racemate, formed from the latter with the same quantity of the *R* enantiomer, will therefore be 16 % and the enantiomeric excess of the *R* product over the racemate will be 84 %. The reaction, as described here using base **A**, has been reported in the literature [9].

❶ 162

In the determination of the most favourable chair conformation of (S)-2-methoxytetrahydropyran the anomeric effect must be considered. This means that a substituent with an unshared electron pair (e.g. a chlorine atom or a methoxy group) in position 2 of a pyran ring prefers to occupy an axial position rather than an equatorial one. The effect can be explained by the repulsion resulting from the dipole-dipole interaction of the unshared electron pairs of the ring oxygen atom with the lone pair electrons of the oxygen atom in the substituent at the anomeric centre. This commonly used simple explanation neglects, however, a vital observation, i.e. that the energy difference between the conformers with substituents in axial and equatorial positions is essentially greater than that which might be expected due to the dipole-dipole repulsion. In actual fact the anomeric effect also causes a shortening of the carbon-oxygen bond. It can be concluded that this bond has partial double bond character which may be explained by hyperconjugation. An unshared electron pair on the ring oxygen atom overlaps with the antibonding orbital of the exocyclic carbon-oxygen bond which is only possible if these two are antiperiplanar to each other. The same effect also operates along the exocyclic C-O bond and as a consequence the methyl group is not antiperiplanar to the C-O bond in the ring.

❶ 163

Non-selective reduction of the ketone groups with LiAlH$_4$ gives rise to a mixture of *cis*- and *trans*-diols. The simplest method to deduce whether the hydrogen atoms in positions 1 and 3 are homotopic, enantiotopic or diastereotopic is the substitution test in which first one, and then the other, hydrogen atom is replaced by another group which is not present in the molecule. An examination of the resulting products from this test leads to the conclusion that the substances are enantiomers and therefore both hydrogen atoms must also be enantiotopic. This result can be confirmed by an inspection of the symmetry of the two diols. In compound **A** there is a centre of symmetry ($i = S_2$) which transforms the two hydrogen atoms into one another. In compound **B** there is a plane of symmetry in which the two hydrogen atoms are reflections of each other.

🛈 164

With the aid of the flow chart in the appendix, the symmetry point group can be determined by a series of questions. Does the compound possess an infinite axis of symmetry? The answer is no. Does the compound possess a finite axis of symmetry and if so, which is the n-fold axis of symmetry of the highest order? The cobalt carbonyl compound has a three-fold axis of symmetry that passes through the apex of the trigonal pyramidal structure. There are no further three-fold axes of symmetry because of the bridging ligands. Are there any C_2 axes perpendicular to the C_3 axis? Again the answer is no. There is no horizontal plane of symmetry but there are three vertical planes of symmetry which each pass through two cobalt atoms and the μ-carbonyl ligands bridging the opposite edge. There is no S_6 axis running parallel to the C_3 axis. Therefore the symmetry point group of $[Co_4(CO)_{12}]$ is C_{3v} and the symmetry elements are C_3 and $3\,\sigma_v$.

🛈 165

The Diels-Alder reaction between $(2E,4Z)$-hexa-2,4-diene with the dieno-phile 2-methoxycyclohexa-2,5-diene-1,4-dione yields four stereoisomers. The Diels-Alder reaction is a stereospecific reaction, therefore the configuration of both the starting materials is relocated in the product. Since the reaction is a concerted [4 + 2] cycloaddition, all the products here have always *cis*-fused rings and because the configuration of $(2E,4Z)$-hexa-2,4-diene remains unaltered, both methyl groups in the product will be *trans* to each other. The various stereoisomers result because the diene and the dienophile may approach each other from both sides and in two distinct orientations. Note that the reaction occurs almost exclusively at the electron poor double bond of the dienophile. In order to deduce the relationships the products have with one another, the configuration of the chirality centres should be determined. Thus **B** and **C** are enantiomers as are **A** and **D**; **A** and **D** are diastereomers of both **B** and **C** since the configuration differs at two chirality centres.

166

Including the stereoisomer shown, the number of possible stereoisomers is $x = 2^7 = 128$ since there are six chirality centres and one double bond which can be either E- or Z-configured. The absolute configuration of the carbon atom at position 4 cannot be determined until the third sphere. In the first sphere there are three carbon atoms, in the second all carbon atoms are attached to two carbon atoms and one hydrogen atom. In the third sphere the priority order is $O,C,C > C,C,(C) > C,C,H$ and therefore this chirality centre is S-configured.

167

From the formula it is obvious that all the six-membered rings are *trans*-fused. This compels them to adopt a chair conformation and hence all the substituents at the ring fusion sites to occupy axial positions. Thus by knowing that the methyl group at position 19 is *trans* to the methoxy group it follows that the methoxy group must occupy an equatorial position.

168

By application of Cram's rule, attack of the Grignard reagent should take place preferentially from the *Si* side since there is less steric hindrance at this side. In order to predict the major product by this rule, the starting material is drawn in the conformation with the carbonyl group antiperiplanar to the largest group on the neighbouring chirality centre, in this case the *tert*-butyl group. After the addition of the propyl group to the structure, the major product is predicted to be (3*S*,4*S*)-2,2,3-trimethylheptan-4-ol.

❗ 169

Calcipotriol contains seven chirality centres and three double bonds where isomerism is possible. Since there are two possibilities for the configuration at each of the stereogenic units, there are in theory $2^7 \cdot 2^3 = 2^{10} = 1024$ possible isomers. In addition, the C6-C7 bond has partial double bond character, thus for every isomer there will be an *s-cis* and an *s-trans* conformer. The configuration of the double bonds, determined from the priority order of the substituents, is indicated in the formula shown below. The absolute configuration at C13 is *R*; the priority order of the attached groups can be determined only in the third sphere. For C14 the digraph for determining the priority order is shown below. Again the priority order can only be determined in the third sphere, in particular in the branch of lowest priority. Here the hydrogen atom of the methyl group has priority over the phantom atom (indicator for non-existent groups) at the duplicate representation of the carbon atom. It is possible to deduce the priority order for C17 in the second sphere, i.e. C(C,C,C) > C(C,C,H) > C(C,H,H) > H. The configuration according to the CIP rules of the chirality centres in the side chains attached to C8 and C17 is given in the formula shown below.

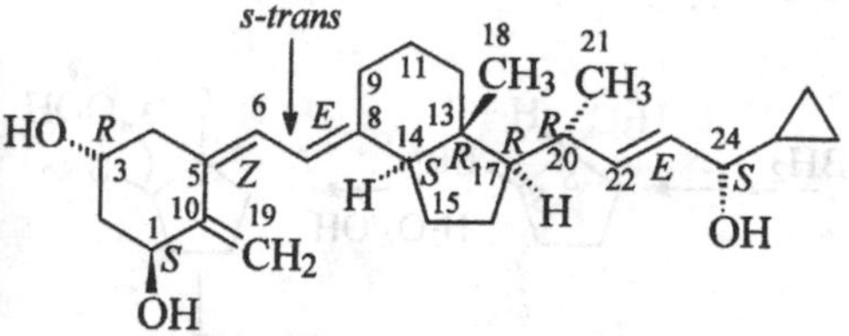

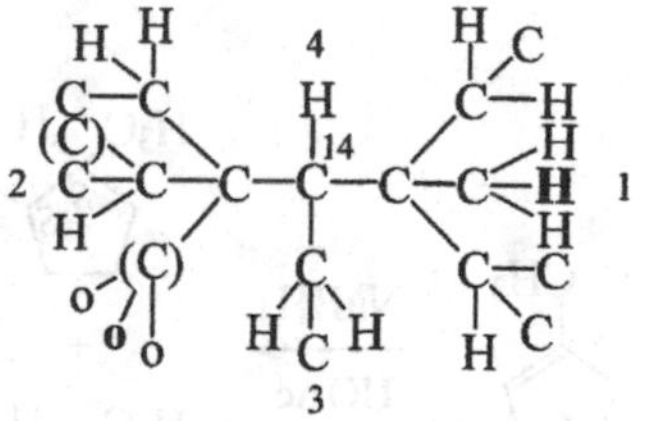

❗ 170

From the reaction conditions given one can easily deduce that hydroboration of the double bond takes place by the addition of the B_2H_6 produced *in situ*. The reaction is a stereospecific *cis* addition reaction in which the hydride ion adds to the more substituted end of the double bond and the boron atom to the other end (anti-Markovnikov addition). Since attack can take place from either side of the double bond, the product (*trans*-2-methylcyclopentyl)borane is a racemate. In the determination of the absolute configuration it should be remembered that a boron atom has a lower priority than a carbon atom when these are attached to the same chirality centre. As the reaction proceeds further addition of the alkene to the remaining B-H bonds occurs to afford a trialkylborane. On adding alkaline hydrogen peroxide to the trialkylborane addition of hydroperoxide anion to the boron atom occurs which is then followed by migration of the alkyl substituent from boron to oxygen with concomitant splitting of a hydroxide ion to afford a trialkyl borate. It is important to note here that the oxidative migration process occurs with retention of configuration. Hydrolysis of the boric ester leads to the racemate of the two possible *trans*-configured alcohols, i.e., (*R*,*R*)- and (*S*,*S*)-2-methylcyclopentanol.

🔴 171

The stereogenic centres are defined using the CIP rules by R and S as well as by r and s depending on whether they are chirality or pseudochirality centres, respectively.

a) The compound has in addition to two chirality centres an s-configured pseudochirality centre. The compound is however, achiral.

b) The compound contains five chirality centres. In order to determine the configuration at the methyl-substituted ring atom it is important to remember a CIP sub-rule that of two constitutionally similar groups the one with an l configuration (i.e. with R,R or S,S configuration) has priority over the group with a u configuration.

c) In addition to the two chirality centres in the side chains, this achiral compound contains a pseudochirality centre in the ring.

d) At position 4 the compound contains a prochirality centre because the absolute configuration of both chirality centres is identical.

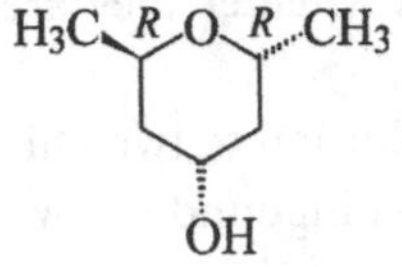

$$H_3C \underset{}{\overset{R}{\diagdown}} O \underset{}{\overset{R}{\diagup}} \cdots CH_3$$

$$OH$$

e) As a result of the priority order of the four groups attached to the quaternary carbon atom the absolute configuration of the molecule shown is S. In this example it is important to note that the carbon atom of the chirality centre is considered as a duplicate representation (i.e. without attached actual substituents) when during the process of determining the priority order the chirality centre itself is encountered as a substituent. Thus the ring segment has preference over the ethoxymethyl side chain – $C((C),H,H > C(H,H,H))$ – whilst the ethoxyethyl side chain has priority over the analogous ring segment since $C(H,H,H) > (C)$. For clarity the digraph required to arrive at the correct stereodescriptor is best considered as two separate digraphs as shown below, i.e. one for the two highest ranking substituents and another for the two substituents of lowest priority.

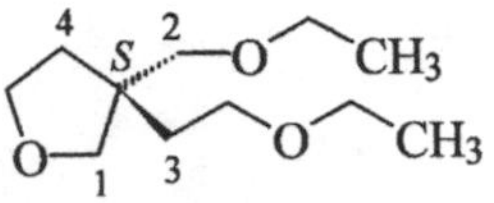

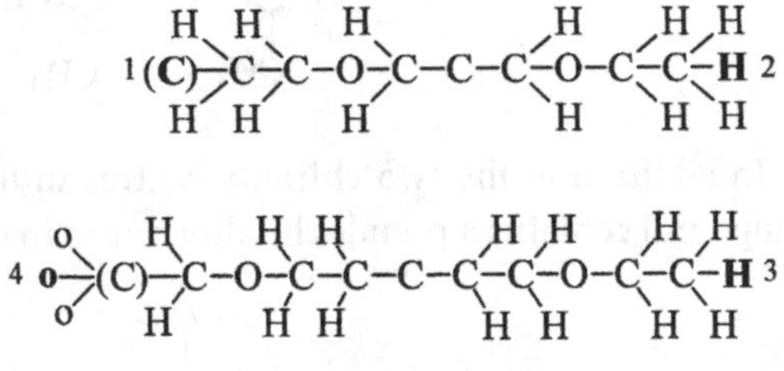

f) The absolute configuration of this coordination compound is denoted by the descriptor A. In order to arrive at this result it is necessary to first of all establish the configuration index. The highest ranking ligand is bromine and this occupies two different sites in the molecule. One of these two atoms lies directly opposite an ammine ligand whilst the other is located *trans* to one of the amino groups of the ethylenediamine ligand. The priority number of the lowest ranking one of these is taken as the first digit of the configuration index. This ligand and the bromine atom thus form the reference axis of the octahedron. If the four ligands in the plane perpendicular to the reference axis are viewed from the direction of the higher ranking atom in the reference axis (the bromine atom) these are clearly seen to be – in the order of decreasing priority – arranged in a counter-clockwise sense. Therefore the descriptor is A.

OC-6-32-A

172

Since the phosphate anion is resonance stabilised, nucleophilic substitution of the bromine atom of the coumarin derivative can occur by either of the two free oxygen atoms. Thus two diastereomers will be produced, which may be distinguished between merely by considering the configuration at the phosphorus atom. In a publication it was shown that using the tetrabutyl-ammonium salt of cyclic AMP (cAMP) in acetonitrile the diastereoisomeric ratio was 85 : 15 in favour of the compound with an S-configured phosphorus atom [10].

🔴 **173**

In order to ascertain which symmetry elements are present in *meso*-tartaric acid, it is necessary to look at the various conformations of the molecule. The symmetrical highest energy conformer, i.e. the synperiplanar conformer (*sp*), has a plane of symmetry in which both enantiomorphic halves of the molecule are reflections of each other. No other symmetry elements are present in this conformation (point group C_s). In the *ap* conformation of *meso*-tartaric acid the only symmetry element present is a centre of symmetry (disregarding the fact that the centre of symmetry is equivalent to any of the infinite number of S_2 axes). The symmetry point group is therefore C_i. All other conformations, e.g. the +synclinal conformation (+*sc*) of *meso*-tartaric acid shown below, are chiral and do not possess any symmetry elements and therefore belong to the point group C_1.

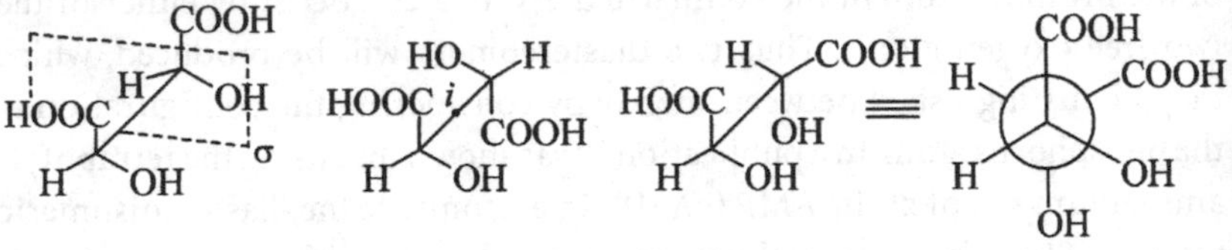

sp   *ap*   +*sc*

🔴 **174**

Memantine is achiral. The molecule has a plane of symmetry in which the enantiomorphic halves of the molecule are reflections of each other. Both the carbon atom attached to the amino group and the tertiary carbon atom are pseudochirality centres and both are *r*-configured. Note that the configuration of the pseudochirality centres remains unaltered on mirror reflection. Although there are four stereogenic centres in the molecule, it is unnecessary to use any stereodescriptor to describe the configuration of the molecule since there are no stereoisomers.

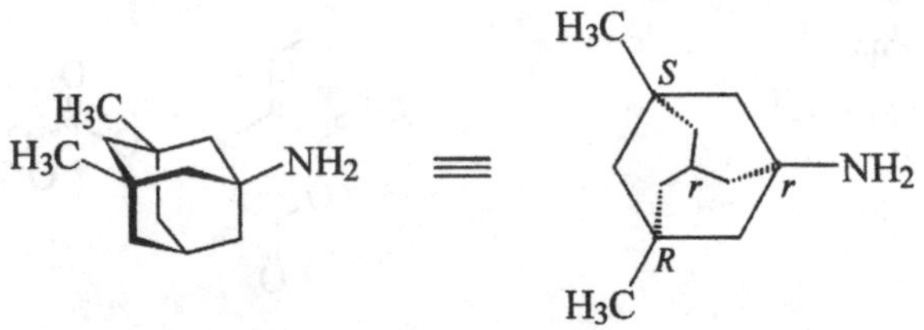

❗ 175

The compound with the ratio of signal intensities 1:1:1 in its ^{1}H NMR spectrum is in accord with structure **A** (point group C_{3v}). The compound with the ratio of signal intensities 2:2:2:1:1:1 in its ^{1}H NMR spectrum is in accord with structure **B** (point group C_s). The chemically equivalent protons are those with the same numbers in the formulae given below. These are homotopic in **A** and enantiotopic in **B**.

A 1:1:1   **B** 2:2:2:1:1:1

❗ 176

Tropatepine exhibits a very interesting structural feature. Normally, compounds with differently configured double bonds are diastereomers of each other. However, in this instance, since two enantiomorphic groups are attached to one end of the double bond, they are in fact enantiomers. It is quite easy to see that both formulae can be transformed into one another by mirror reflection. Tropatepine is therefore chiral. It is used clinically as the racemate.

❗ 177

a) B_5H_9 has a square pyramidal structure. The compound belongs to the symmetry point group C_{4v}. It has one C_4 axis and four planes of symmetry σ_v two of which pass through opposite corners and two of which bisect opposite edges of the square plane.

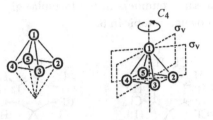

b) B_4H_{10} belongs to the symmetry point group C_{2v}. It has one C_2 axis and two planes of symmetry σ_v.

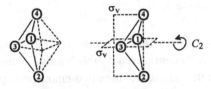

c) B_6H_{10} has a pentagonal pyramidal structure. The compound belongs to the symmetry point group C_{5v}. It has one C_5 axis of symmetry and five vertical planes of symmetry σ_v whose line of intersection is the C_5 axis.

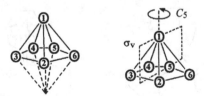

d) B_5H_{11} possesses only a plane of symmetry σ and therefore belongs to the symmetry point group C_s.

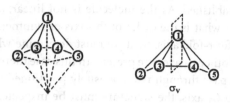

❶ 178

From the priority order of the groups attached to the sulfur atom (F > N > chlorophenyl > phenyl) the starting material is S-configured. On the assumption that an S_N2 reaction at the sulfur atom also proceeds with Walden inversion, the reaction intermediate will have a trigonal bipyramidal structure with the fluorine atom and hydroxy group occupying apical positions. After tautomerisation of the initially formed product, the resulting sulfoximine will have an R configuration. Such a mechanism is inferred from the results of a detailed kinetic investigation reported in reference [11].

❗ 179

By means of the flow chart given in the appendix the symmetry point group can be easily established. As the molecule is not linear, the first question to be answered is what is the order of the axis of symmetry of the highest order? This is a four-fold axis. Next we must determine whether this is the only C_4 axis or if other C_4 axes are present. In this case there are two other C_4 axes which each pass through two oppositely positioned palladium atoms. Since there is no C_5 axis, the structure must be inspected to see whether there are any C_3 axes. It is now meaningful to look at the chlorine atoms in the molecule. A total of four three-fold axes of symmetry pass through the middle of triangles formed from three chlorine atoms or three palladium atoms, respectively. Since the question of four-fold axes has already been answered, it only remains to see whether a centre of symmetry is present. This is the case and therefore $[(PdCl_2)_6]$ has the symmetry point group O_h. There are, in addition to the symmetry elements already established above, six C_2 axes passing through opposite pairs of chlorine atoms. There are also three planes of symmetry each containing four palladium atoms and six planes of symmetry diagonal to these planes each containing two palladium atoms and two chlorine atoms. There are also three S_4 axes coinciding with the C_4 axes and four S_6 axes coinciding with the C_3 axes. Note that the palladium atoms are located at the corners of an octahedron the edges of which have a bridging chlorine atom.

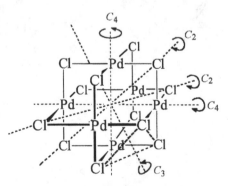

❶ 180

Doxycycline contains six chirality centres whose configuration is indicated in the structural formula shown below. The chirality centre at position 12a requires careful consideration since the priority order of the groups attached to this atom may be dependent upon which tautomer is present. As an illustration, if one of the possible tautomers is considered (right hand formula) then the priority order can only be established in the fourth sphere as indicated in the digraph depicted below.

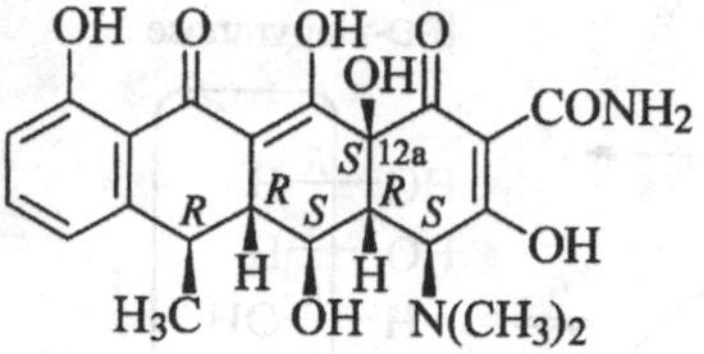

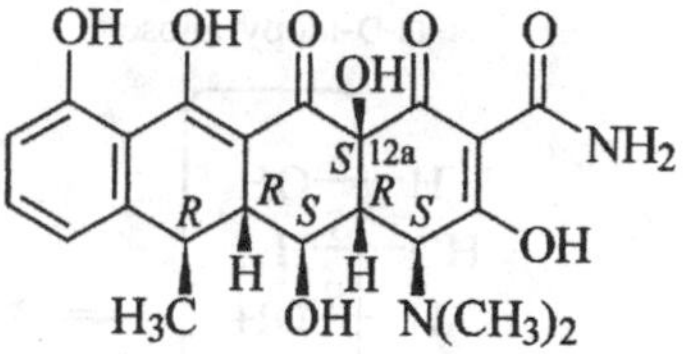

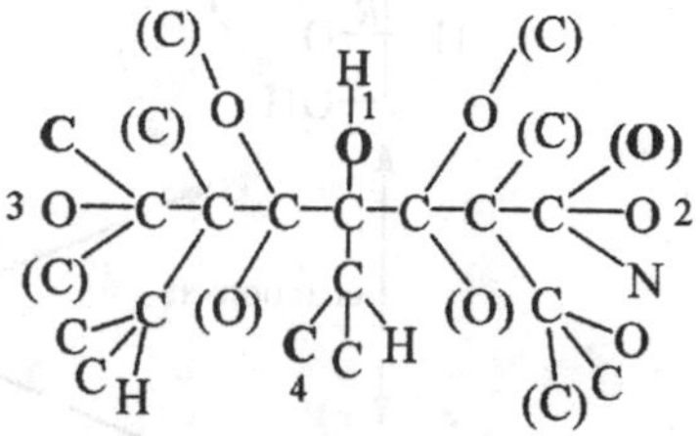

❗ 181

α-L-Idopyranose is the enantiomer of α-D-idopyranose. Similarly, β-L-ido-pyranose is the enantiomer of β-D-idopyranose. α-D-Idopyranose is an epimer of β-D-idopyranose since they are distinguishable solely in the absolute configuration at C1. α-L-Idopyranose and β-L-idopyranose are similarly epimers. α-D-Idopyranose and β-L-idopyranose are diastereomers of each other since they have different absolute configuration at C2, C3, C4 and C5. Likewise β-D-idopyranose and α-L-idopyranose are diastereomers.

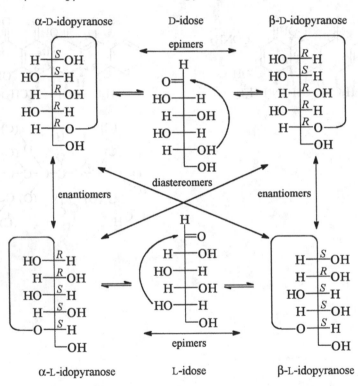

❗ 182

The complex can exist with both a *cis* and a *trans* configuration. Since the compound has two bidentate ethylenediamine ligands, the isomers with both chlorine atoms in the *cis* configuration will exist as a pair of enantiomers. These will have the descriptors *OC*-6-2'2-Δ and *OC*-6-2'2-Λ, respectively, and belong to the symmetry point group C_2. The *trans* isomer has a plane of symmetry and three mutually perpendicular C_2 axes and therefore belongs to the point group D_{2h} and has the descriptor OC-6-12'.

trans   *cis*   *cis*

Δ   Λ

🛈 183

Osazones are bis-phenylhydrazones of carbohydrates formed when aldoses are treated with an excess of phenylhydrazine. During the course of the reaction the hydroxy group at position 2 is oxidised to an oxo group (with simultaneous release of aniline and ammonia). Subsequent reaction of the intermediate ketone with phenylhydrazine yields the osazone. Therefore sugars which are epimers at position 2 form the same osazone. Similarly, 2-ketoses can produce this osazone. In this case oxidation of the primary hydroxy group (adjacent to the original keto group) occurs during the reaction. The osazone shown can be obtained from D-galactose, D-talose and D-tagatose irrespective of whether these are the α or β anomers since the reaction proceeds via the open chain tautomer of the sugar.

β-D-galactopyranose   α-D-galactopyranose   β-D-talopyranose

α-D-talopyranose   α-D-tagatopyranose   β-D-tagatopyranose

184

Because the double bond has an E configuration, the total number of configurational isomers which are theoretically possible is $2^4 = 16$ (four chirality centres). However, as a result of constitutional symmetry twelve of the theoretically possible isomers are in fact six pairs of identical compounds. Therefore only ten stereoisomers are possible:

$1R,1'R,2R,2'R$;
$1R,1'R,2S,2'S$;
$1R,1'R,2R,2'S \equiv 1R,1'R,2S,2'R$;
$1S,1'S,2R,2'R$;
$1S,1'S,2S,2'S$;
$1R,1'S,2R,2'R \equiv 1S,1'R,2R,2'R$;
$1R,1'S,2S,2'S \equiv 1S,1'R,2S,2'S$;
$1R,1'S,2S,2'S \equiv 1S,1'R,2S,2'R$;
$1R,1'S,2S,2'R \equiv 1S,1'R,2R,2'S$;
$1S,1'S,2R,2'S \equiv 1S,1'S,2S,2'R$.

According to the pharmaceutical literature however, both positions 1 and 1' have an R configuration in the finished drug and hence only the first three of the isomers listed above are components. The formula of the second isomer in the list is shown below.

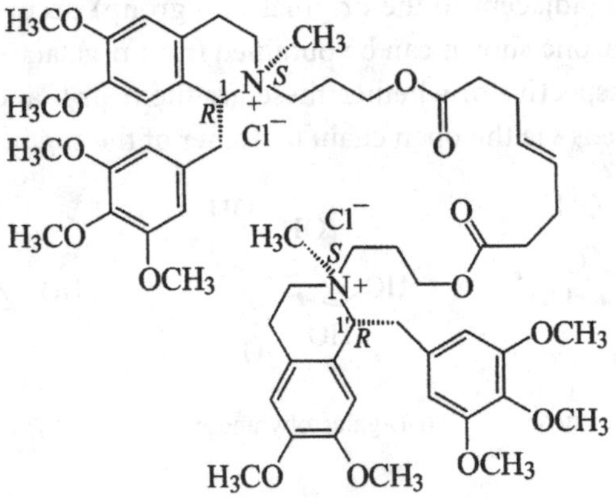

185

On the basis of the idealised structure depicted, copper(I) benzoate belongs to the symmetry point group D_{2d}. The symmetry elements present are three C_2 axes, an S_4 axis and two dihedral planes of symmetry σ_d (these are referred to as dihedral because they are vertical planes of symmetry which each bisect the angle between two of the horizontal C_2 axes). The result can be checked by referring to the flow chart given in the appendix. In reality the copper atoms in this compound are not arranged in a square but in the form of a parallelogram. Moreover, the benzoate groups are not actually perpendicular to the plane containing the four copper atoms [12].

❶ 186

The compound has two chirality centres and three pseudochirality centres. There is however, only one (achiral) diastereomer of the compound shown in the question. The two isomers can be distinguished from one another solely on the relative position of the chlorine or bromine atoms which lie in a plane which also happens to be the plane of symmetry of the molecule (this is the only symmetry element present, therefore the symmetry point group is C_s). It is possible in this instance to specify the configuration unequivocally using the descriptors E and Z. However, in systematic nomenclature the complete configuration of all the stereogenic centres is specified. Thus the (so-called) "Z" isomer is ($1s,3r,5R,6r,7S$)-1,6-dibromo-3,6-dichloroadamantane and the "E" isomer is ($1s,3r,5R,6s,7S$)-1,6-dibromo-3,6-dichloroadamantane, i.e. the two isomers can be distinguished simply by the descriptor used for position 6.

❗ 187

Vancomycin contains eighteen chirality centres: nine in the carbohydrate side chain and nine in the aglycone moiety. The configuration of all these chirality centres is shown in the structural formula given below. In addition the biphenyl group in the aglycone also contains an S_a-configured chirality axis and there are two chirality planes. Determination of the configuration is possible since rotation of the chlorine substituted benzene rings at room temperature is highly hindered. The pilot atoms required for the determination of the configuration of the chirality planes are marked in the formula with stars. The sugar carbohydrate side chain is an α-L-vancosamine substituted β-D-glucopyranosyl unit. Details regarding the structural properties, strategies for the total synthesis [13] and biosynthesis [14] of vancomycin and analogous glycopeptide antibiotics can be found in the latest reviews.

❶ 188

In order to be able to decide which transformations must be carried out at which groups in the starting molecule it is first of all best to convert the Fischer projection formula of the target molecule to a zigzag projection which corresponds to the orientation of the substituents in the starting material. Once this has been accomplished it is a relatively simple task to identify the necessary reaction steps.

Thus benzylation of the hydroxy group with retention of the configuration and subsequent oxidative cleavage of the double bond, e.g., by ozonolysis followed by a reductive work-up is the most obvious entry. (Dihydroxylation of the double bond followed by cleavage of the intermediate glycol is also a valid alternative to the ozonolysis step. Subsequent reduction of the product would be required in this case, too.) The integrity of all chirality centres is unaffected by these steps. However, since both the newly formed hydroxymethyl groups have higher priority in the CIP system than a nitrogen substituted carbon atom, the stereodescriptors at both outer chirality centres will change. After removal of the *tert*-butyldimethylsilyl (TBDMS) group with tetrabutylammonium fluoride (TBAF) – also with retention of the configuration – the priority order of the groups at the central chirality centre is reversed. Conversion of the 1,2-diol moiety to its acetonide proceeds with retention but leads to a further change in the priority order at this chirality centre. Finally, Swern oxidation of the primary hydroxy group to yield the desired aldehyde once more changes the priority order of groups at the central chirality centre. (Details regarding the synthesis of **B** outlined here can be found in the original literature [15].)

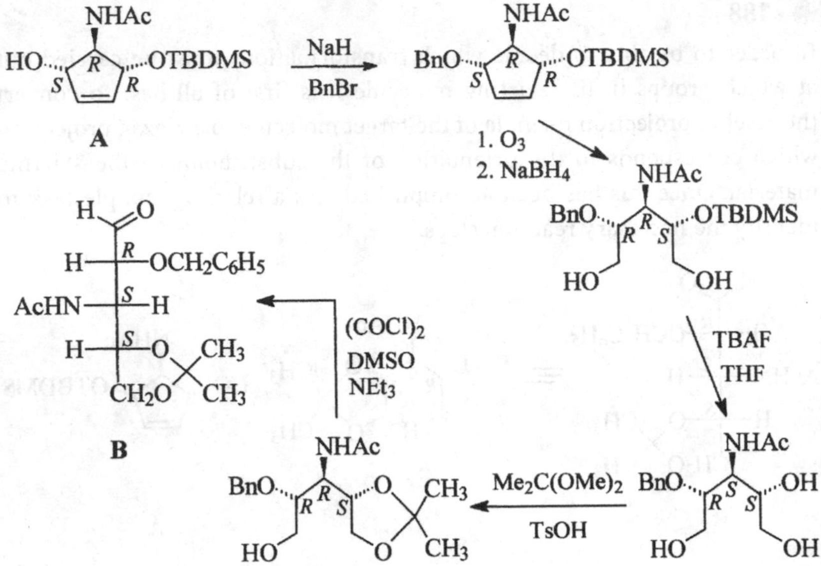

NHAc

HO⋯⟨R⟩⋯OTBDMS →(NaH / BnBr)→ BnO⋯⟨R⟩⋯OTBDMS

S   R   S   R

A

1. O₃
2. NaBH₄

NHAc

BnO⋯⟨R⟩⟨S⟩⋯OTBDMS

HO   OH

H   O

H—⟨R⟩—OCH₂C₆H₅

AcHN—⟨S⟩—H

H—⟨S⟩—O   CH₃

CH₂O   CH₃

B

←(COCl)₂ / DMSO / NEt₃←

NHAc

BnO⋯⟨R⟩⟨S⟩—O   CH₃

HO   O   CH₃

←(Me₂C(OMe)₂ / TsOH)←

NHAc

BnO⋯⟨R⟩⟨S⟩⋯OH

HO   OH

(TBAF / THF)

🔴 189

There are two possible strategies for devising a synthesis of methyloxirane. The first possibility is the disconnection approach, i.e., to analyse the target molecule for what a potential precursor might look like. At first glance direct epoxidation of propene in the presence of a chiral catalyst might appear attractive. However, handling of gaseous propene is problematic and propene does not contain any directing groups. Moreover, methyloxirane is reactive and very volatile (b.p. 34 °C) and both these factors would make separation of the enantiomers difficult. Taking into account the above noted problems identified in the direct approach a reaction involving ring closure must now be considered. The requirement here is to have an intermediate alcohol with a leaving group on a vicinal carbon atom. If the alcohol is a primary alcohol then a ring closure step involving displacement of the leaving group at position 2 will proceed with inversion whilst if the alcohol is secondary a similar reaction step will proceed with retention of the configuration.

The second approach is to look for a suitable naturally occurring compound with a three carbon chain and a chirality centre as the starting material. The obvious candidates are lactic acid and alanine or one of their derivatives. The scheme shown on the right outlines two of the possible synthetic routes starting from (S)-lactic acid ethyl ester, which is both cheap and commercially available, as the precursor to both the desired enantiomers based on the synthetic strategy discussed above. Mesylation of the hydroxy group followed by reduction of the ester function affords the primary alcohol. Treatment of this alcohol with base proceeds with inversion to yield (R)-methyloxirane. Protection of the hydroxy group in the original ester as its tetrahydropyran-2-yl (THP) ether using 3,4-dihydro-2H-pyran (DHP), followed by reduction of the ester yields the primary alcohol which can then be transformed into a leaving group by tosylation. Removal of the THP ether by treatment with acid furnishes a secondary alcohol which cyclises in the presence of base with retention of configuration to give (S)-methyloxirane.

🔵 190

Tranylcypromine contains a cyclopropane ring and construction of this unit must be the overriding consideration in planning a synthesis for the compound. On the basis of the constitution given for the compound a ring closure reaction appears to be rather impracticable. Consequently, cyclo-addition reactions must be considered for which three possibilities exist as shown by the disconnections a, b, or c in the formula below.

The relative configuration of both substituents (the amino group and the phenyl ring) appears at first sight to indicate that a is the most attractive option. This approach corresponds to the addition of carbene generated from diiodomethane and zinc to a carbon-carbon double bond (Simmons-Smith reaction). The enamine required in this approach is not readily accessible and therefore (E)-1-nitro-2-phenylethene, which is easily prepared by an aldol condensation reaction, would have to be used. Unfortunately, the Simmons-Smith reaction does not work with electron deficient double bonds. For broadly similar reasons approach b can be discounted. The remaining possibility c raises the question whether a compound containing a suitable substituent which can be transformed subsequently to an amino group can be added to styrene (B). The most appropriate compound for this purpose is ethyl diazoacetate (G). On heating this ester nitrogen is lost with the formation of a carbene which adds to styrene to yield a mixture of trans- and cis-2-phenylcyclopropanecarboxylic acid ethyl ester (rac-C and rac-D) in a 65 : 35 ratio. The amount of the thermodynamically more stable rac-C in the mixture can be increased to 95 % by epimerisation with sodium ethanolate. The remaining cis isomer can be removed after hydrolysis of the ester group by recrystallisation. The carboxylic acid (rac-E) can then be converted into its amide or its acyl azide and converted to the amine by a degradation reaction (the Hofmann degradation or the Curtius degradation, respectively). Both these reactions involve a rearrangement to yield an intermediate isocyanate and during this process the configuration of the migrating group (the cyclopropyl ring in this example) is unaltered, i.e., the *trans*-acid derivative yields the *trans*-amine (*rac*-**A**).

$$B + G \longrightarrow rac\text{-}C + rac\text{-}D \longrightarrow rac\text{-}E + rac\text{-}F$$

$$rac\text{-}E \longrightarrow \longrightarrow rac\text{-}A$$

❶ 191

Propranolol contains a 1,2,3-trisubstituted propane unit which can easily be obtained by the ring opening of an epoxide by a nucleophile, isopropylamine in this instance. The *S*-configured epoxide **C** is an essential intermediate in the synthesis and this compound can be envisaged as being derived from allyl alcohol. This alcohol can be converted by Sharpless epoxidation depending upon the choice of the chiral reagent employed, (+)- or (−)-diisopropyl tartrate (DIPT), to either (*R*)-oxiranylmethanol (**A**) or its enantiomer, *ent*-**A**. All synthetic steps proceeding from **A** must be chosen carefully to avoid changing the configuration of the chirality centre. This can be achieved by tosylation of the hydroxy group followed by substitution of the tosylate by 1-naphthol to yield the desired epoxide **C**. It is also possible to prepare the *S* enantiomer of propranolol from *ent*-**A** if the naphthyloxy group is introduced at the other end of the three carbon chain or if the synthetic sequence contains a step which involves inversion at the chirality centre. One such possibility is outlined in the scheme shown overleaf where the epoxide is first reacted with sodium 1-naptholate. Treatment of the resulting diol with HBr in glacial acetic acid proceeds with concomitant selective esterification of the secondary hydroxy group and substitution of the primary hydroxy group. Subsequent alkaline hydrolysis of the latter affords an alkoxide which undergoes direct ring closure to give epoxide C.

Appendix

Selected substituent groups listed in the order of increasing priority according to the CIP system

dimethoxyboryl
methyl
ethyl
propyl
butyl
pentyl
hexyl
isopentyl
isobutyl
allyl (prop-2-enyl)
neopentyl
prop-2-ynyl
benzyl
4-chlorobenzyl
isopropyl
vinyl (ethenyl)
sec-butyl
cyclopropyl
cyclobutyl
cyclopentyl
cyclohexyl
prop-1-enyl
tert-butyl
isopropenyl
ethynyl
phenyl
4-(dihydroxyboryl)phenyl
p-tolyl
4-nitrophenyl
4-methoxyphenyl
m-tolyl
3,5-dimethylphenyl
3-nitrophenyl
3,5-dinitrophenyl
prop-1-ynyl
o-tolyl
2,6-dimethylphenyl
mesityl (2,4,6-trimethylphenyl)
trityl (triphenylmethyl)
2-nitrophenyl
2,4-dinitrophenyl
aminomethyl
hydroxymethyl
formyl
acetyl

propanoyl
benzoyl
carboxy
methoxycarbonyl
ethoxycarbonyl
benzyloxycarbonyl
tert-butoxycarbonyl
amino
methylamino
ethylamino
benzylamino
isopropylamino
tert-butylamino
phenylamino
acetylamino
benzoylamino
(benzyloxycarbonyl)amino
(*tert*-butoxycarbonyl)amino
dimethylamino
diethylamino
dipropylamino
piperidino
morpholino
phenyldiazenyl
nitroso
nitro
hydroxy
methoxy
ethoxy
benzyloxy
phenoxy
acetoxy
benzoyloxy
mesyloxy (methylsulfonyloxy)
tosyloxy [(4-methylphenyl)sulfonyloxy]
fluoro
dimethyl(phenyl)silyl
diphenylphosphanyl
sulfanyl
methylsulfanyl
methylsulfinyl
mesyl (methylsulfonyl)
chloro
bromo
iodo

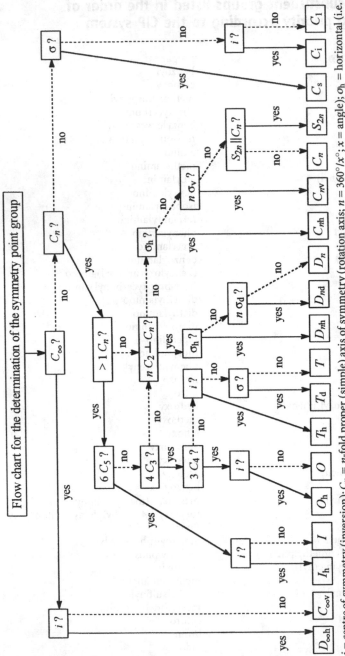

Flow chart for the determination of the symmetry point group

i = centre of symmetry (inversion); C_n = n-fold proper (simple) axis of symmetry (rotation axis; $n = 360°/x°$; x = angle); σ_h = horizontal (i.e. perpendicular to C_n) plane of symmetry (mirror plane); σ_v = vertical (i.e. parallel with C_n) plane of symmetry; σ_d = dihedral (i.e. bisecting the angle between C_2 axes) plane of symmetry; S_{2n} = alternating (improper) axis of symmetry (rotation-reflection); n always means the order of the proper axis of symmetry of the highest order in the molecule.

Bibliography

1. Text books and introductory works

Ernest L. Eliel, Samuel H. Wilen: *Stereochemistry of Organic Compounds*, Wiley, New York, Chichester, Brisbane, Singapore, Toronto, 1994

E. L. Eliel, S. H. Wilen: *Organische Stereochemie* (shortened translation), Wiley-VCH, Weinheim, New York, Chichester, Brisbane, Singapore, Toronto, 1997

Ernest L. Eliel, Samuel H. Wilen, Michael P. Doyle: *Basic Organic Stereochemistry*, Wiley, New York, Chichester, Weinheim, Brisbane, Singapore, Toronto, 2001

Sheila R. Buxton, Stanley M. Roberts: *Guide to Organic Stereochemistry from methane to macromolecules*, Prentice Hall/Pearson Education Limited, Harlow, 1996

Karl-Heinz Hellwich: *Stereochemie – Grundbegriffe*, Springer-Verlag, Berlin, Heidelberg, New York, 2001; English edition in preparation

David G. Morris, *Stereochemistry*, The Royal Society of Chemistry, Cambridge, 2001

Gerhard Quinkert, Ernst Egert, Christian Griesinger: *Aspects of Organic Chemistry, Structure*, Verlag Helvetica Chimica Acta, VCH, Basel, 1996

Michael J. T. Robinson, Organic Stereochemistry, Oxford University Press, Oxford, 2000

Bernard Testa: *Principles of Organic Stereochemistry*, Marcel Dekker, New York, Basel, 1979

Alexander von Zelewsky: *Stereochemistry of Coordination Compounds*, John Wiley & Sons, Chichester, 1996

Siegfried Hauptmann, Gerhard Mann: *Stereochemie*, Spektrum Akademischer Verlag, Heidelberg, 1996

Hermann J. Roth, Christa E. Müller, Gerd Folkers: *Stereochemie & Arzneistoffe*, Wissenschaftliche Verlagsgesellschaft, Stuttgart, 1998

Christoph Rücker, Joachim Braun: *UNIMOLIS, A Computer-aided Course on Molecular Symmetry and Isomerism*, http://unimolis.uni-bayreuth.de

2. Further reading
a) Cited publications

[1]   Hui-Ping Guan, Yao-Ling Qiu, Mohamad B. Ksebati, Earl R. Kern, Jiri Zemlicka: *Synthesis of phosphonate derivatives of methylenecyclopropane nucleoside analogues by alkylation-elimination method and unusual opening of cyclopropane ring*, Tetrahedron **58**, 6047–6059 (2002)

[2]   H.-P. Buchstaller, C. D. Siebert, R. H. Lyssy, G. Ecker, M. Krug, M. L. Berger, R. Gottschlich, C. R. Noe: *Thieno[2,3-b]pyridinones as Antagonists on the Glycine*

Site of the N-methyl-D-aspartate Receptor – Binding Studies, Molecular Modeling and Structure-Activity-Relationships, Sci. Pharm. **68**, 3–14 (2000)

[3]   F. Albert Cotton, Jan M. Troup: *Accurate Determination of a Classic Structure in the Metal Carbonyl Field: Nonacarbonyldi-iron*, J. Chem. Soc. Dalton Trans. **1974**, 800–802

[4]   Jeffrey A. Robl, Chong-Qing Sun, Jay Stevenson, Denis E. Ryono, Ligaya M. Simpkins, Maria P. Cimarusti, Tamara Dejneka, William A. Slusarchyk, Sam Chao, Leslie Stratton, Ray N. Misra, Mark S. Bednarz, Magdi M. Asaad, Hong Son Cheung, Benoni E. Abboa-Offei, Patricia L. Smith, Parker D. Mathers, Maxine Fox, Thomas R. Schaeffer, Andrea A. Seymour, Nick C. Trippodo: *Dual Metalloprotease Inhibitors: Mercaptoacetyl-Based Fused Heterocyclic Dipeptide Mimetics as Inhibitors of Angiotensin-converting Enzyme and Neutral Endopeptidase*, J. Med. Chem. **40**, 1570–1577 (1997)

[5]   Robin A. Fairhurst, Steven P. Collingwood, David Lambert, Elke Wissler: *Nucleic Acid Containing 3'-C-P-N-5' Ethyl Phosphonamidate Ester and 2'-Methoxy Modifications in Combination; Synthesis and Hybridisation Properties*, Synlett **2002**(5), 763–766

[6]   Daniel Guillon, Michael A. Osipov, Stéphane Méry, Michel Siffert, Jean-François Nicoud, Cyril Bourgogne, P. Sebastião: *Synclinic-anticlinic phase transition in tilted organosiloxane liquid crystals*, J. Mater. Chem. **11**(11), 2700–2708 (2001)

[7]   Jesse Dambacher, Mikael Bergdahl: *Employing the simple monosilylcopper reagent, Li[PhMe₂SiCuI], in 1,4-addition reactions*, Chem. Commun. **2003**, 144–145

[8]   Xun-Wei Wu, Xue-Long Hou, Li-Xin Dai, Ju Tao, Bo-Xun Cao, Jie Sun: *Synthesis of Novel N,O-planar chiral [2,2]paracyclophane ligands and their application as catalysts in the addition of diethylzinc to aldehydes*, Tetrahedron Asymmetry **12**, 529–532 (2001)

[9]   Manfred Braun, Brigitte Meyer, Boris Féaux de Lacroix: *Synthesis of (R)- and (S)-O-Methylcannabispirenone by Desymmetrization of O-Methylcannabispirone*, Eur. J. Org. Chem. **2002**, 1424–1428

[10] Torsten Eckardt, Volker Hagen, Björn Schade, Reinhardt Schmidt, Claude Schweitzer, Jürgen Bendig: *Deactivation Behaviour and Excited-State Properties of (Coumarin-4-yl)methyl Derivatives. 2. Photocleavage of Selected (Coumarin-4-yl)methyl-Caged Adenosine Cyclic 3',5'-Monophosphates with Fluorescence Enhancement*, J. Org. Chem. **67**(3), 703–710 (2002)

[11] Tiaoling Dong, Takayoshi Fujii, Satoro Murotani, Huagang Dai, Shin Ono, Hiroyuki Morita, Choichiro Shimasaki, Toshiaki Yoshimura: *Kinetic Investigation on the Hydrolysis of Aryl(fluoro)(phenyl)-λ^6-sulfanenitriles*, Bull. Chem. Soc. Jpn. **74**, 945–954 (2001)

[12] Michael G. B. Drew, Dennis A. Edwards, Roger Richards: *Crystal and Molecular Structure of Tetrakis[copper(I) benzoate]*, J. Chem. Soc. Dalton Trans. **1977**, 299–303

[13] K. C. Nicolaou, Christopher N. C. Boddy, Stefan Bräse, Nicolas Winssinger: *Chemistry, Biology, and Medicine of the Glycopeptide Antibiotics*, Angew. Chem. **111**(15), 2230–2287 (1999), Angew. Chem. Int. Ed. **38**(15), 2096–2152 (1999)

[14] Brian K. Hubbard, Christopher T. Walsh: *Vancomycin Assembly: Nature's Way*, Angew. Chem. **115**(7), 752–789 (2003), Angew. Chem. Int. Ed. **42**(7), 730–765 (2003)

[15] Haiyan Lu, Zhuoyi Su, Ling Song, Patrick S. Mariano: *A Novel Approach to the Synthesis of Amino-Sugars. Routes To Selectively Protected 3-Amino-3-deoxy-aldopentoses Based on Pyridinium Salt Photochemistry*, J. Org. Chem. **67**, 3525–3528 (2002)

b) IUPAC rules and recommendations

* *Basic Terminology of Stereochemistry*, Pure Appl. Chem. **68**(12), 2193–2222 (1996)

* *Nomenclature of Carbohydrates*, Pure Appl. Chem. **68**(10), 1919–2008 (1996)

* *The Nomenclature of Lipids, Recommendations 1976*, Eur. J. Biochem. **79**, 11–21 (1977)

* *Nomenclature and Symbolism for Amino Acids and Peptides (Recommendations 1983)*, Pure Appl. Chem. **56**(5), 595–624 (1984); Eur. J. Biochem. **138**, 9–37 (1984)

Graphical Representation of Configuration, Pure Appl. Chem., in press

International Union of Pure and Applied Chemistry (IUPAC), Organic Chemistry Division, Commission on Nomenclature of Organic Chemistry, J. Rigaudy, S. P. Klesney, Eds.: *Nomenclature of Organic Chemistry, Sections A, B, C, D, E, F and H, 1979 Edition*, Pergamon Press, Oxford, 1979

International Union of Pure and Applied Chemistry, Organic Chemistry Division, Commission on Nomenclature of Organic Chemistry (III.1): *A Guide to IUPAC Nomenclature of Organic Compounds, Recommendations 1993*, Blackwell Scientific Publications, Oxford, 1993

* *Corrections to A Guide to IUPAC Nomenclature of Organic Compounds (IUPAC Recommendations 1993)*, Pure Appl. Chem. **71**(7), 1327–1330 (1999)

International Union of Pure and Applied Chemistry: *Nomenclature of Inorganic Chemistry – IUPAC Recommendations 2005*, International Union of Pure and Applied Chemistry/The Royal Society of Chemistry, Cambridge, 2005

Those references above marked with an asterisk (*) can also be accessed via the internet at the address http://www.chem.qmul.ac.uk/iupac/.

c) Literature on special topics

R. S. Cahn, Sir Christopher Ingold, V. Prelog: *Specification of Molecular Chirality*, Angew. Chem. **78**, 413–447 (1966), Angew. Chem. Int. Ed. Engl. **5**, 385–415 + 511 (1966)

Vladimir Prelog, Günter Helmchen: *Basic Priciples of the CIP-System and Proposals for a Revision*, Angew. Chem. **94**, 614–631 (1982), Angew. Chem. Int. Ed. Engl. **21**, 567–583 (1982)

Günter Helmchen: *Nomenclature and Vocabulary of Organic Stereochemistry*, in: Methods of Organic Chemistry (Houben-Weyl), Volume E21a, Stereoselective Synthesis, Thieme, Stuttgart, New York, 1995, pp. 1–74

Dieter Seebach, Vladimir Prelog: *The Unambiguous Specification of the Steric Course of Asymmetric Syntheses*, Angew. Chem. **94**, 696–702 (1982), Angew. Chem. Int. Ed. Engl. **21**, 654–660 (1982)

Robert S. Ward: *Selectivity in Organic Synthesis*, Wiley, Chichester, New York, Weinheim, Brisbane, Singapore, Toronto, 1999

István Hargittai, Magdolna Hargittai: *Symmetry through the Eyes of a Chemist*, VCH, Weinheim, 1986

Jan M. Fleischer, Alan J. Gushurst, William L. Jorgensen: *Computer Assisted Mechanistic Evaluations of Organic Reactions. 26. Diastereoselective Additions: Cram's Rule*, J. Org. Chem. **60**(3), 490–498 (1995)

Peter R. Schreiner: *Teaching the Right Reasons: Lessons from the Mistaken Origin of the Rotational Barrier in Ethane*, Angew. Chem. **114**(19), 3729–3731 (2002), Angew. Chem. Int. Ed. **41**(19), 3579–3581 (2002)

Benito Alcaide, Pedro Almendros: *The Direct Catalytic Asymmetric Cross-Aldol Reaction of Aldehydes*, Angew. Chem. **115**(8), 884–886 (2003), Angew. Chem. Int. Ed. **42**(8), 858–860 (2003)

Frieder W. Lichtenthaler: *Emil Fischer's Proof of the Configuration of Sugars: A Centennial Tribute*, Angew. Chem. **104**(12), 1577–1593 (1992), Angew. Chem. Int. Ed. Engl. **31**(12), 1541–1556 (1992)

Lucette Duhamel, Pierre Duhamel, Jean-Christophe Plaquevent: *Enantioselective protonations: fundamental insights and new concepts*, Tetrahedron Asymmetry **15**(23), 3653–3691 (2004)

Fritz Vögtle, Joachim Franke, Arno Aigner, Detlev Worsch: *Die Cramsche Regel*, Chem. unserer Zeit **18**(6), 203–210 (1984)

Index

The references given in this index are the numbers of the questions and answers, not page numbers.

The International Nonproprietary Names (INN) or proposed INNs (pINN) of those drugs discussed in this book are indicated accordingly in the index.